CALIFORNIA NATURAL HISTORY GUIDES

INTRODUCTION TO
TREES OF THE
SAN FRANCISCO BAY REGION

Introduction to

TREES of the
SAN FRANCISCO
BAY REGION

Glenn Keator

UNIVERSITY OF CALIFORNIA PRESS

California Natural History Guide Series No. 65

University of California Press
Berkeley and Los Angeles, California

University of California Press, Ltd.
London, England

© 2002 by the Regents of the University of California

Library of Congress Cataloging-in-Publication Data

Keator, Glenn.
    Introduction to trees of the San Francisco Bay Region / Glenn Keator.
        p.   cm. — (California natural history guides ; 65)
    Includes bibliographical references and index.
    ISBN 0-520-23005-1 (hardcover : alk. paper) — ISBN 0-520-23007-8 (pbk. :
alk. paper)
    1. Trees—California—San Francisco Bay Area—Identification.   2. Trees—
California—San Francisco Bay Area—Pictorial works.   I. Title. II. Series.

QK149 .K35   2003
582.16′09794′6—dc21

                                                                          2002029140

Manufactured in China
10   09   08   07   06   05   04   03   02   01
10   9   8   7   6   5   4   3   2   1

The paper used in this publication meets the minimum requirements of
ANSI/NISO Z39.48-1992 (R 1997) (*Permanence of Paper*). ⊚

The publisher gratefully acknowledges the generous
contributions to this book provided by

the Moore Family Foundation
Richard & Rhoda Goldman Fund
and
the General Endowment Fund of the University of
California Press Associates.

CONTENTS

Trees are the first plants we notice in a landscape. They define spaces by dominating them, whether as parts of home gardens, as specimens in parks, in rows along streets, or as forests and woodlands in the "wild." Trees are simply the largest living things that we encounter wherever we go. Naturally, trees are the first elements of the landscape we want to know about: their names, their growth patterns, their flowers and fruits, their sizes, their life cycles.

This book has been written to answer those questions and address other concerns. Because the San Francisco Bay Region enjoys a mild Mediterranean climate, which allows us to grow an enormous number of different kinds of trees, this book is limited to naturally occurring trees—trees that grow on their own in the "wild" and in reserves and natural parklands. The book does not attempt to describe what you see in people's gardens or in planned public parks; the trees described here are either native or introduced species that grow naturally on their own outside gardens and tended parks. Nor does this book include every single kind of tree seen in the wild, because it's simply not possible to be exhaustively comprehensive. For example, the reader may stumble across a deodar cedar *(Cedrus deodara)* in some coastal locale or encounter a normally shrubby coffee berry specimen *(Rhamnus californica)* that has developed a treelike trunk and so qualifies for tree status.

Although the San Francisco Bay Region usually includes the nine counties surrounding San Francisco Bay, I'm adding three important adjacent counties: Santa Cruz and Monterey in the south and Mendocino in the north. Altogether, this book covers Monterey, Santa Cruz, San Mateo, San Francisco, Santa Clara, Alameda, Contra Costa, Marin, Solano, Napa, Sonoma, and Mendocino counties.

This book can be approached in more than one way. If you already know the name of a tree you're curious about, look it up in the index or find it among the species accounts—entries are alphabetized by scientific name, although you'll need to look up the family first in the table of contents. If you're simply looking at the illustrations in order to peruse the possibilities, leaf through the species accounts, nearly all of which have one or more photos.

If, on the other hand, you're trying to make a positive identification of a tree you don't know, turn to the key. If you're unfamiliar with keys, see "How to Use a Key," below.

What Is a Tree?

Nature seldom has hard-and-fast rules, and this is certainly true for trees. A tree is usually defined as a woody plant that grows more than 10 feet tall and bears one or a few major trunks from which smaller limbs and branches diverge. But the line that separates trees from shrubs—woody plants with multiple main stems and growing less than 10 feet tall—is tenuous. For this reason, many large shrubs that occasionally exhibit a dominant trunk structure are included here. Environment may also determine whether a particular plant grows as a shrub or a tree.

Names of Trees

It's the name we turn to first when we want to learn more about a particular tree. You'll find that trees bear two kinds of names: trivial, or common, names that are in everyday use by the average person; and scientific, or Latin, names that are used throughout the horticultural and botanical world. There are advantages and disadvantages to both kinds of names, but for greater precision, scientific names are preferred. Common names are not always standardized, and many trees—for example, *Quercus chrysolepis*, aka goldcup oak, canyon live oak, or maul oak—have more than one common name. Common names may also allude to relationships that do not exist. The

Douglas-fir *(Pseudotsuga menziesii)* is not a true fir because the details of its bark, needles, and cones are considerably different from those of the genus *Abies,* though it does belong to the pine family (Pinaceae) along with true firs. Sometimes a common name may lead you even farther astray. The Russian-olive *(Elaeagnus angustifolia)* is not even remotely related to the true olive *(Olea europea);* it belongs to the oleaster family (Elaeagnaceae), whereas the olive belongs to its own family, Oleaceae. When common names such as Douglas-fir become so embedded in the language that they are a permanent fixture, the name is hyphenated to indicate that it represents a special combination. So, for instance, we have tanbark-oak *(Lithocarpus densiflorus)* for an oaklike tree that is not a true oak (i.e., not one of the *Quercus* species), and incense-cedar *(Calocedrus decurrens)* for a tree with fragrant wood that is not a true cedar (i.e., not one of the *Cedrus* species).

Not every tree has a well-established common name, but because people seem to want a common name for every tree, I have provided one. Not everyone will agree with my choice for such common names, and some will disagree with my use of one common name over another. Where a tree has more than one well-known common name, I have tried to indicate that.

Scientific names are based on carefully crafted rules for naming and are in a latinized form. Because scientific names can be recognized by scientists throughout the world regardless of the language spoken, they give a real sense of permanence and are immediately recognizable anywhere. Each scientific name consists of at least two parts: a genus name, given first and capitalized; and a specific epithet, given second and starting with a lowercase letter. (Think of how most people use two names to identify themselves.) Both genus name and specific epithet are italicized. Although the initial reaction of the novice may be that scientific names are impossibly "difficult," in fact, English speakers have a real advantage, because many of the latinized names have cognates in English. Here are a few examples of recognizable scientific names. *Pinus radiata* (Monterey pine): the genus name is the Latin word for pine, and the specific epithet is a word that means the cones are in whorls or radiate out along the branch. *Olea europea* (olive): the genus name is derived from the Latin word for oil (think of our word *oleo*) because olive oil has been such an important commodity, and the specific epithet tells us the tree comes from Europe. *Phoenix canariensis* (Canary Island date palm): the genus name indicates a tree that seems to grow in difficult places, as the mythical bird, the phoenix, rises out of its ashes; the specific epithet tells us this tree is native to the Canary Islands. And so it goes; every scientific name has a story.

It's also important to understand that scientific names may change as more is learned about a particular tree. For example, when the coast redwood *(Sequoia sempervirens)* and the giant sequoia *(Sequoiadendron giganteum)* were first described, it was widely believed that they both belonged to the genus *Sequoia*. But subsequent studies showed that the giant sequoia was sufficiently different from the coast redwood that it should be placed in its own genus. When such a situation arises, the first species named *Sequoia* is retained in the genus, and the second species must be given a new genus name.

It's also important in the scheme of naming and classifying to know the hierarchy used. Throughout this book, I refer to *family, genus,* and *species.* Families are the largest category described here; they are assemblages of related plants that share certain key features. For this reason, before describing several related trees, I start with an overview of the family they belong to. Some families are relatively small; others are large, with thousands of species. Family names always end in *-aceae,* for example, Rosaceae (rose family), Pinaceae (pine family), Cupressaceae (cypress family), Oleaceae (olive family).

A genus (plural: *genera*) consists of closely related kinds of trees: oaks (genus *Quercus*), pines (genus *Pinus*), firs (genus

Abies), cypresses (genus *Cupressus).* The kinds of oaks or pines are what we call *species* (singular and plural are the same). Some genera have only one species; others have dozens or even hundreds. The Bay Area is blessed with many species of oaks, including the coast live oak *(Q. agrifolia),* blue oak *(Q. douglasii),* valley oak *(Q. lobata),* and several more.

Some species are so variable that they are subdivided into subspecies or varieties. Some botanists make a distinction between subspecies and variety, others do not. Whenever we encounter a variety or subspecies, we indicate the name preceded by the abbreviation *var.,* for variety, or *subsp.,* for subspecies. For instance, the beach pine *(Pinus contorta)* has been subdivided into at least three subspecies: *P. contorta* subsp. *contorta* (the regular beach pine), *P. contorta* subsp. *bolanderi* (the Bolander pine), and *P. contorta* subsp. *murrayana* (the lodgepole pine), the latter found in high mountains. Note that the subspecies designation follows the specific epithet, and that the name is italicized and starts with a lowercase letter.

Major Attributes of Trees

There are many kinds of trees in the world, and so there are many ways of sorting them. Some attributes of trees immediately call attention to themselves; other features may go unnoticed until you take a closer look or already know what to look for. Outstanding features include the tree's ultimate size and shape, leaf characteristics, bark patterns, and flowers, cones, or fruits. Let's examine each category in turn.

Size and Shape

Many trees are immediately singled out for their size, yet size can be misleading—especially height. The age of the tree and its environment play major roles in determining size. For the novice, it may be difficult to know whether a particular tree is a sapling or a full-grown specimen, although if you scrutinize the area where many of the same trees are present, the tree's age is usually obvious. Many young trees exhibit forms of foliage and overall shape that differ strikingly from their mature forms. Environment has an equally dramatic effect: strong winds, poor soils, and little water may severely limit the size of a tree. For example, the pygmy cypress *(Cupressus goveniana* subsp. *pygmaea)* is a stunted runt of a tree when it lives on the shallow, nutrient-deficient soils of ancient marine terraces. The same tree will grow to over 100 feet high on deep, rich soils. For these reasons, the overall size of a given tree may be a poor guide to identification.

Shape, by contrast, is often more reliable, at least when the approximate age of the tree is known. For example, a young incense-cedar exhibits a narrow, conical shape, whereas an old tree may bear an irregular, craggy crown. Many mature trees are quickly identified by the overall shape of their crown, often from some distance away. Practice makes perfect in applying this knowledge. Look for whether the crown is broader than tall, rounded, drooping, narrow, cone-shaped, irregular, or symmetrical.

Leaf Characteristics

When available, leaves are among the easiest means of identifying trees. Trees can be conveniently divided according to the major details of their leaves. A convenient starting point, for example, may be whether a tree is deciduous or evergreen. Most deciduous trees lose all their leaves in fall and remain leafless through the winter. Even in spring and summer, deciduous trees have rather thin, pliable leaves. Evergreen trees keep at least one or more sets of leaves through an entire year, so that they always have some leaves, even in winter. But no evergreen tree keeps all of its leaves for its entire life; the typical tenure of an evergreen leaf is between two and four years. Leaves of evergreen trees are often tough and leathery.

Another major criterion is whether leaves are needle- or scalelike. The majority of trees that bear seed cones rather than fruits—the conifers—have such leaves. By contrast, the majority of flowering trees have broader leaves, leading to the commonly used term *broadleaf tree*. Note that the term *hardwood tree* is synonymous with the term *broadleaf tree*. All softwood trees are conifers. The terms *softwood* and *hardwood* are not absolute: for example, balsa belongs to the hardwood group but has very soft, pliable wood. Rather, these terms are applied in a broad manner: flowering trees are hardwood, and coniferous trees are softwood.

Once you're past these major categories, many subtler leaf details readily separate one kind of tree from another. Here is a list of attributes to look for:

Leaf Arrangement

Are leaves found in pairs, whorls (three or more at a node), or singly on their twigs? Paired or whorled leaves are much less common than single (alternate) leaves.

Leaf Form

Are leaves simple, consisting of one piece, or are they compound, with two or more separate leaflets? Simple leaves may be further distinguished by whether they're deeply lobed or not.

Compound leaves are sometimes confused with a branch carrying several simple leaves. How can you tell the difference? Look for a tiny bud at the base of the leaf: a compound leaf has only one bud at its very base, and none at the base of each leaflet. If the leaf also has stipules (pairs of leaflike appendages) at its base, they will be found only at the base of the whole leaf, not the base of the leaflets.

Compound leaves are either pinnate—leaflets arranged ladderlike along a central stalk—or palmate—leaflets arrayed like fingers on a hand. Pinnately compound leaves are much more common than palmately compound leaves; but in general, there are far more trees with simple leaves than compound ones.

Leaf Shape

There are almost as many leaf shapes as there are kinds of leaves. The names of many shapes are intuitive, for example, oval, linear, lanceolate, and elliptical. Others are easily deduced: orbicular (round), for the shape of an orb; deltoid (triangular), for the shape of a river's delta. If in doubt, look up the meaning in the glossary in the back of this book. Some leaves may combine shapes of two different patterns: ovate-lanceolate, for example, which is a shape halfway between ovate and lanceolate.

Leaf Margin

The edges of leaves may be smooth (entire) or lined with jagged teeth or rounded scallops, or be wavy or curled under. There are many useful terms for describing these attributes.

Leaf Apex and Base

Some leaves with fundamentally the same shape may differ in the details of their tips. Leaf tips may be narrowly or broadly rounded or squared off, or may taper gradually or abruptly to a point.

Vein Pattern

Leaf veins often form intricate patterns, and each pattern can uniquely identify a particular tree. For example, close scrutiny of vein patterns may reveal that the primary pattern is pinnate, whereas the secondary pattern (the smaller veins) is reticulate. Often, leaves with the same overall shape have differing vein patterns, making this feature important in separating look-alikes.

Leaf Color

Although most leaves are green, there are many different tones of green. Some leaves also have a decided grayish, reddish, or bluish cast.

Leaf Coverings

Although it's not always obvious, many leaves are covered with special glands or hairs—protective structures that keep the leaves from drying out or from being chewed by insects. A good 10× hand lens is a useful adjunct for revealing details that to the untrained eye seem obscure. As you learn more about these minute coverings, it also becomes obvious that there are many different kinds of hairs—stiff, short, long, silky, straight, curled, starburstlike, interlaced, and more.

Leaf Size

Just as the overall size of a tree may be misleading, so, too, is leaf size. Typically, leaves at the top of a tree are smaller than leaves on the bottom branches. Shade leaves are generally larger because they need as broad a surface as possible to trap sunlight; sun leaves are smaller in order to conserve water.

Leaf Age

As trees grow to maturity, they lose their juvenile leaves, and the juvenile leaves on young trees may look entirely different from the adult leaves of the same tree. For example, blue gum has broadly oval, opposite juvenile leaves and vertically oriented, alternate, sickle-shaped adult leaves.

Bark Patterns

The bark of mature trees has its own distinctive pattern, but this pattern may be different or not be obvious when the tree is young. Tree bark may be brown, gray, whitish, reddish, or other colors. Bark texture may be stiff and tough, soft and spongy, or corky and flaky. Because old bark peels away every year and each tree has its own pattern of sloughing off its bark, bark shows distinctive designs. In some trees the bark is ribbed or ridged vertically; in others, it is crosshatched or checkered; in still others it is flaky or puzzlelike. Fissures between the raised areas of bark may be very shallow or well over an inch deep. Different layers of bark may exhibit different colors; for example, the western sycamore's *(Platanus racemosa)* oldest outer bark is gray, whereas its youngest inner bark is cream colored. Unfortunately, it is often difficult to describe the complexities of bark patterns in a few words: the eye is able to record these differences much more faithfully than language can.

Flowers, Cones, or Fruits

Broadleaf, or hardwood, trees usually produce flowers that ripen into fruits containing seeds. Coniferous, or softwood, trees usually bear cones—tiny pollen cones in spring, and seed cones that ripen into conspicuous papery, fleshy, or woody cones later in the year. Whether the tree is a conifer or has flowers, its reproductive parts are the most reliable means of identification, and botanists rely heavily on the characteristics of cones, flowers, and fruits for classifying trees. The biggest problem is that flowers and fruits are usually ephemeral—they may last only a few days or weeks. The more substantial fruits, such as acorns and samaras, may persist after they've fallen from the tree's branches. The same is true for the seed cones of conifers. But because bark and leaves are present most of the year, the keys in this book rely on them more than on the reproductive parts.

There are two general categories of broadleaf (flowering) trees: those with tiny, inconspicuous, wind-pollinated flowers, and those with showy, insect-pollinated flowers. In general, most of our native flowering trees belong to the first category. For example, oaks produce hanging chains (catkins) of tiny male flowers that turn yellow when they shed their pollen.

These male flowers have no colorful petals to attract insects. Oak female flowers are even less conspicuous—they're about the size of a pea and occur in the axils of new leaves. Each female flower is green and topped by three tiny red stigmas.

Generally, only the smaller trees, such as the California buckeye *(Aesculus californica)*, California fremontia *(Fremontodendron californicum)*, western redbud *(Cercis occidentalis)*, and service berries (*Amelanchier* spp.) produce colorful flowers with petals that are aimed at drawing bees, butterflies, or birds for pollination. Some of these, like the flowers of many of our native shrubs, are counted among the showiest and most beautiful components of native landscapes.

Successfully fertilized flowers (except exclusively male flowers) are followed by fruits—ripened ovaries that contain one or more seeds. Mature fruits may become fleshy and be brightly colored to attract mammals and birds, they may turn brown and papery and split open to shed their seeds, or they may harden into brown, thick-walled nuts, acorns, or small achenes (one-seeded fruits that remain closed) that animals gather for food and bury for later use. Throughout this book, you'll discover examples of each of these distinctive categories.

Conifers, by contrast, produce pollen and seed cones. Seed cones may take anywhere from six months to over two years to ripen and reach full size. Mature cones may be as short as an inch or as long as two feet in the case of the sugar pine. Typically, they consist of whorled or spirally arranged papery or woody scales. Each scale carries two or more seeds that, when ripe, generally fall out of the cone. Wings on the seeds help in dispersing the seeds on winds. A few are wingless and dispersed by animals.

Tree Habitats

In many places, trees dominate over smaller plants, and consequently, many ecologists accept the idea that trees represent the climax vegetation for any given area. But there are many mitigating circumstances, especially in California: coastal bluffs buffeted by constant wind; the unstable sandy soils of shifting sand dunes; constantly wet, boggy places; areas that are frequently burned; or dry habitats with nutrient-poor soils. To be sure, trees have adapted to some of these difficult situations, but there are definitely habitats where tree growth is untenable. One of the most dramatic of these is chaparral —a dense cover of evergreen shrubs adapted to steep, rocky, hot slopes that receive small amounts of annual precipitation. Human practices have also rendered many areas devoid of trees through burning, clearing, grazing, and plowing.

Despite these treeless places, there are many habitats that favor trees. A survey of Bay Area plant communities counts no fewer than seven kinds that support trees. Starting at the immediate coast, we find dense forests of closed-cone pines and cypresses clinging to rocky soils on promontories that receive heavy summer fog and ample winter rain. These forests consist of one or more species of wind-adapted conifers whose seed cones normally open only after fire sweeps through. These cones remain permanently attached to the trees. Typical of this community are Monterey pine *(Pinus radiata)*, bishop pine *(P. muricata)*, Monterey cypress *(Cupressus macrocarpa)*, and Santa Cruz cypress *(C. abramsiana)*. An unusual and extreme version of this vegetation occurs on ancient marine terraces whose soils have long been leached of nutrients and that have developed an underlying hardpan: the trees here are stunted. Several sites on the Mendocino coast are typified by pygmy cypress *(C. goveniana* subsp. *pygmaea)*, bishop pines, and Bolander's beach pines *(P. contorta* subsp. *bolanderi)*, none of which grow up to more than 10 feet tall on such sites.

A second kind of closed-cone pine forest can be found on inland, nutrient-poor soils where summer days often top 100 degrees F. Typical of the unusual soils that support these trees are serpentinites, developed from the slick bluish or reddish rocks that are toxically high in heavy metals. Here, look for knobcone pine *(P. attenuata)*, Sargent cypress *(C. sargentii)*, and Macnab cypress *(C. macnabiana)*.

Redwood forest is arguably California's most famous tree habitat. Although people typically think of redwoods as hugging the coast, they seldom occur closer than a half mile to the ocean because of their sensitivity to wind (redwood roots are shallow) and salt spray. Redwood forests reach their best development on flood plains within the heart of the summer fog belt. In old-growth forests, the coast redwood *(Sequoia sempervirens)* shades out other trees, although on slopes or in second-growth forests, the tanbark-oak *(Lithocarpus densiflorus)*, madrone *(Arbutus menziesii)*, Douglas-fir *(Pseudotsuga menziesii)*, California bay *(Umbellularia californica)*, and other trees mix with the redwoods.

These latter trees and others—mostly broadleaf evergreen trees—continue farther inland in a forest of their own called mixed-evergreen forest. Other trees to be expected in this community include the California-nutmeg *(Torreya californica)*, coast live oak *(Quercus agrifolia)*, and coast chinquapin *(Chrysolepis chrysophylla)*. Past fire history, duration or lack of coastal fogs, and elevation influence the mix of any given stand of this complex forest.

Mixed-evergreen forest often subtly merges with a higher-elevation forest that is uncommon in the Bay Area: mixed-conifer forest (aka ponderosa pine forest). This forest is extensive in California's mountainous regions, often occurring between 2,500 and 6,000 feet elevation. Look for patches of it on the Bay Area's highest mountains, especially on Mt. St. Helena in Napa County, and on Mt. Hamilton in Santa Clara County. It is much more characteristic in the rugged Santa Lucia Mountains of Monterey County. The dominant trees of this forest include ponderosa and sugar pines *(P. ponderosa* and *P. lambertiana)* and incense-cedar *(Calocedrus decurrens)*, as well as some of the trees from the mixed-evergreen forest, especially Douglas-fir. Broadleaf trees are also present where the taller conifers have not shaded them out and may include large stands of California black oak *(Q. kelloggii)*, goldcup oak *(Q. chrysolepis)*, and localized populations of flowering dogwood *(Cornus nuttallii)*.

Inland at lower elevations and beyond the main fog belt, mixed-evergreen forest merges into a more open woodland dominated by oaks or oaks mixed with gray pine *(P. sabiniana)* and California buckeye *(Aesculus californica)*. In many places, oak woodlands dominate on summer-hot, south-facing slopes, whereas mixed-evergreen forests cling to cooler, north-facing slopes.

Each area of oak woodland has its own mix of oaks, but the hottest, driest places are typified by blue oak *(Q. douglasii)* and interior live oak *(Q. wislizenii);* valley bottoms by valley oak *(Q. lobata)* and coast live oak *(Q. agrifolia);* places with higher rainfall by Garry oak *(Q. garryana)* and California black oak.

Finally, permanent water courses support their own version of a treeland referred to as riparian woodland. This exuberant and fast-growing plant community thrives because of its constant and reliable source of water, but it is normally only a few trees broad, and only large flood plains (parts of the Russian River, for example) have well-developed riparian areas. Within this woodland the trees grow in distinct layers and are often festooned by vines. The upper story may be overshadowed by Fremont cottonwood *(Populus fremontii)*, western sycamore *(Platanus racemosa)*, bigleaf maple *(Acer macrophyllum)*, box elder *(A. negundo)*, red and white alders *(Alnus rubra* and *A. rhombifolia)*, and Oregon ash *(Fraxinus latifolia);* the lower strata feature one or more kinds of willow *(Salix* spp.), elderberries *(Sambucus* spp.), and a number of deciduous shrubs. Each area that supports riparian woodland has a different mix of these deciduous trees; usually only two or three species dominate any given place.

One last tree community deserves special comment: areas that have been planted with blue gum eucalyptus *(Eucalyptus globulus)* have maintained themselves and have sometimes expanded. Because of the intense competition of blue gum roots for water and because of the inhibitory nature of the oils produced by their leaves, few other species—woody or not—occur in these impoverished woodlands.

Threats to Tree Habitats

As urbanization expands, natural tree habitats are lost. But it's not just houses and their attendant roads, parking lots, and shopping malls that threaten trees. Air pollution seems to be an inevitable by-product of urbanization, and a number of conifers and other trees are adversely affected by it. Another of the myriad threats, and one of the most seemingly innocuous, is agriculture: monocultural stands of vegetables, fruit trees, and pastures for cows and sheep. Huge tracts of woodlands and forests have been altered and eliminated in order to grow food and raise livestock.

Other less obvious causes are also of concern. One is the alteration of our fauna; extirpation of most large predators, especially mountain lions, has led to overpopulation by deer. Deer browsing of young saplings becomes a major problem when old and dying trees are not replaced. The stress of drought—and we're now well aware of our drought cycles—exacerbates the situation: deer become so desperate for food that they'll even resort to supposedly deer-proof plants such as the California bay (*Umbellularia californica*).

Yet another cause for worry is our sometimes accidental meddling with our flora. As noxious weeds have made their way into pastures, roadsides, and gardens so, too, aggressive nonnative trees have taken root in a variety of habitats. Some of these intruders—the English holly (*Ilex aquifolium*), maytens (*Maytenus boaria*), and plume albizia (*Albizia jubrissin*), for example—appear to cover a relatively small percentage of natural territory; however, even these have the potential to dominate an area by means of vigorous stump sprouting and suckering.

Other trees have invaded ecosystems in a much more aggressive manner and threaten not only the native trees but the habitats they create for other plants and animals. The blue gum eucalyptus *(E. globulus)* is a famous example, which is often taken for granted as native by many Bay Area residents. Not only are the blue gum's roots thirsty, but the leaves also carry on chemical warfare and suppress the growth of other plants. Another culprit is the salt-cedar, or tamarisk (*Tamarix* spp.), which threatens water courses by thirsty, deeply probing roots that not only use the water that native shrubs and trees would otherwise have used but also reduce the overall water flow of the stream itself.

Finally, some introduced trees can hybridize with our native trees. For example, *Crataegus monogyna* (a European hawthorn) is closely enough related to our native and rather uncommon western hawthorn *(C. suksdorfii)* that they hybridize. The resulting hybrids have the potential to be both invasive and better adapted to their potential home than either species by itself, and the gene pool of the native hawthorn may become seriously altered.

There is no easy solution to these and other dilemmas that affect native trees, but whatever measures can be taken to slow the process of habitat loss are important to maintaining the integrity of our unique tree communities. Let us remain aware of the problems and ready to do battle on behalf of the trees, which have no obvious defenses of their own. Our wild lands deserve to remain wild and diverse.

How to Use a Key

Although photographs promise easy identification, they are not always reliable. Seldom does any one image reveal enough details to make an identification certain. Hence the need for keys. Although keys may seem challenging, they can also be fun.

A key is a logical device that allows the reader to progressively narrow the choices until only one species remains. Each step in a key has two choices; in this book they're numbered similarly (1a versus 1b, for example). Each choice talks about one or two traits that can be compared—say, for example, whether flowers are red or yellow. After making the first choice, go to the next step and another pair of choices. You repeat this process until you've arrived at the name of your particular plant. Of course, if all choices were as easy as noting flower color, keys would be a breeze and everyone would use them without any special effort or practice. Sadly, there are many reasons why keys prove challenging and difficult.

It is of great importance to remember that nature is variable. Although a given species might usually have red flowers, it is not uncommon for flower color to vary, and yellow variants sometimes occur. No keys are comprehensive enough to take all possible exceptions into account. So, a first rule in keying is to look at more than one plant to see if your particular specimen is typical.

Second, keys often contain human errors. Inconsistencies may creep in even when the person making the key has been careful. This sort of problem may be hard to detect at first, but one check against taking a wrong turn—regardless of whose error it is—is to read the description and look at the photo(s) for the tree you've identified. If these don't match, something has gone wrong during the keying process.

Third, measuring things requires careful attention. I've tried to minimize the use of size in my keys. Remember, for example, when measuring leaves to choose a typical leaf. Always measure the broadest part of the leaf when width is given. Use only fresh leaves or flowers for measurements; dried material shrinks considerably.

Fourth, there are many specialized terms that are needed to describe important facets of leaves, twigs, bark, fruits, and flowers. Make frequent use of the glossary when you start. I've tried to minimize the use of technical terms, but sometimes a specific word for a specialized part is essential.

Finally, make careful notes about the tree you're identifying, especially if you don't have this book in hand while you're out in the field. What is the overall shape of the tree? Is it evergreen or deciduous? What are the leaves like? Does it have any obvious flowers, cones, or fruits? If so, what are the details? What color and pattern is the bark? Is there anything special about the branching pattern? What is the habitat like: shady, moist, dry, sunny, coastal, inland? What are some other obvious trees or shrubs growing with the tree? Is the tree growing on any especially distinctive soils such as those derived from blue serpentinite rocks?

Two aids that you may find useful to help identify trees, especially in the field, are a hand lens and a pair of binoculars. A good quality 10× hand lens is best, for it magnifies enough to reveal the details of hairs and other tiny structures, yet doesn't magnify so much that the image is dim. (The greater the magnification, the more light is required.) Binoculars are useful for seeing leaf patterns and other features of trees from a distance. You can also reverse the binoculars and use them as a sort of hand lens for seeing close-up details.

Here are some hints to help you get started with the keying process:

- Always start at the beginning and jot down the choices you make. If you come to a step where you're unsure of the choice, make a note of it.
- A wrong choice usually leads you to other choices that contradict what you see or are obviously wrong in some respect.
- Try working a key backward by starting with a tree whose name you already know. This will help give you insight into how the keys in this book work and will show you what I had in mind when I created the key.

- Always check the species accounts after making your identification. Helpful clues may be found there, such as where the tree occurs or the other trees that grow with it.
- Always check the photo(s) of the tree you've identified. If there is a discrepancy between the way the tree appears in the photograph and what you remember seeing, be suspicious that something went wrong with your choices during keying. (Also bear in mind that things such as tree shape may vary.)

Good luck. The more you practice with keys, the easier the process becomes.

DISTINGUISHING FEATURES OF TREES

LEAVES

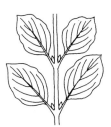

Opposite,
petiolate

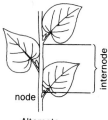

node
internode

Alternate

Whorled

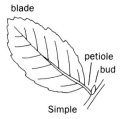

blade
petiole
bud

Simple

once
pinnate

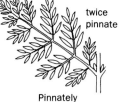

twice
pinnate

Pinnately
compound

Palmately
compound

LEAF SHAPES

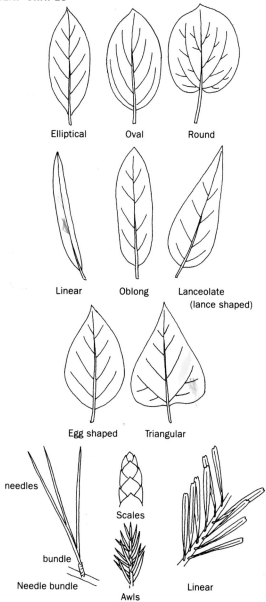

Elliptical

Oval

Round

Linear

Oblong

Lanceolate
(lance shaped)

Egg shaped

Triangular

needles

bundle

Needle bundle

Scales

Awls

Linear

LEAF MARGINS

Entire

Scalloped

Serrated

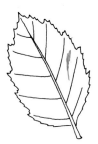

Doubly serrated

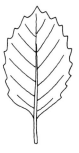

Toothed

Spiny

Pinnately lobed

Palmately lobed

TWIGS

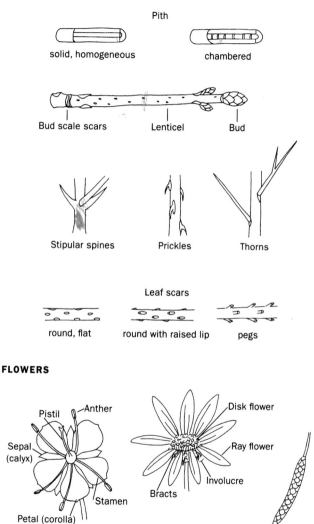

Pith

solid, homogeneous   chambered

Bud scale scars   Lenticel   Bud

Stipular spines   Prickles   Thorns

Leaf scars

round, flat   round with raised lip   pegs

FLOWERS

Pistil   Anther

Sepal (calyx)

Petal (corolla)

Stamen

Disk flower

Ray flower

Involucre

Bracts

Male catkin

FRUITS AND CONES

Samara

Achene

Follicle

Legume

Pome

Drupe

Nut

Berry

Capsule

Utricle

Aril

Seed-bearing
catkin

Seed-bearing
cone

KEY TO BAY AREA TREES

1a Leaves needle-, awl-, or scalelike; seeds usually borne in cones 2

1b Leaves broader; seeds borne in fruits . . . . preceded by flowers 21

    2a Mature leaves scalelike, borne in twos or threes. . . 3

    2b Mature leaves needle- or awl-like, borne singly or in bunches on spur shoots. 6

3a Pink flowers appear in spring; plants live in wet places . . . . tamarisk, salt-cedar (*Tamarix* spp.)

3b Woody or fleshy seed cones in place of flowers; plants don't live in wet places. 4

    4a Branches distinctly flattened; outer cone scales resembling birds' wings in flight . . . . incense-cedar *(Calocedrus decurrens)*

    4b Branches three dimensional; outer cone scales not as above. 5

5a Seed cones hard and woody, seldom opening except on hot days or after fire . . . . cypresses (*Cupressus* spp.)

5b Seed cones fleshy and berrylike . . . . California juniper *(Juniperus californica)*

    6a Needles borne in bunches of two to five on tiny spur shoots. 7

    6b Needles borne singly and directly along twigs. . . 14

7a Seed cones remain on trees indefinitely and don't fall . . 8

7b Seed cones fall from tree soon after they ripen 10

    8a Needles in threes on spur shoots. 9

    8b Needles usually in twos on spur shoots . . . . bishop pine (*Pinus muricata*)

9a Trees scrawny and often multitrunked; seed cones display prominent knobby spines . . . . knobcone pine (*Pinus attenuata*)

9b Trees usually symmetrical with a full crown, seldom

multitrunked; seed cones not prominently spiny.......
.... Monterey pine *(Pinus radiata)*

10a Needles on spur shoots in twos or fives......... 11

10b Needles on spur shoots in threes............... 12

11a Needles in twos; seed cones not more than two-and-a-half inches long .... beach pine *(Pinus contorta)*

11b Needles in fives; seed cones usually more than a foot long
.... sugar pine *(Pinus lambertiana)*

12a Needles gray or gray green; seed cones massive and heavy 13

12b Needles deep green; seed cones light, seldom more than four inches long ....
.... ponderosa pine *(Pinus ponderosa)*

13a Needles wispy and drooping; seed cones about as broad as long.................... gray pine *(Pinus sabiniana)*

13b Needles bushy and not drooping; seed cones longer than broad.................... Coulter pine *(Pinus coulteri)*

14a Needles shed attached to twigs (look at the leaf duff under the trees). 15

14b Needles shed individually. 16

15a Needles on lower branches arranged flat in two rows along twigs; seed cones no more than two inches long ..
.... coast redwood *(Sequoia sempervirens)*

15b Needles on lower branches spirally arranged; seed cones average three inches long ....
.... giant sequoia *(Sequoiadendron giganteum)*

16a Needles glossy and tipped with a daggerlike spine
.... 17

16b Needles not both glossy and tipped with a sharp spine.. 18

17a Needles not prominently white striped underneath; seed cones fleshy and plumlike when ripe..................
.... California-nutmeg *(Torreya californica)*

17b Needles prominently white striped underneath; seed cones composed of papery scales; cones shatter when ripe.................. Santa Lucia fir *(Abies bracteata)*

18a Needles don't leave behind pegs or rough areas

when they fall; seed cones sit upright on branches .... grand fir *(Abies grandis)*

18b Needles leave behind pegs or rough areas when they fall; seed cones hang......................... 19

19a Needles leave distinct pegs when they fall; seed cone scales are papery and often fluted.....................
.... Sitka spruce *(Picea sitchensis)*

19b Needles leave indistinct rough places when they fall; seed cones scales are not fluted......................... 20

20a Seed cones have prominent three-pronged bracts between scales; leaders are stiff and straight ....
.... Douglas-fir *(Pseudotsuga menziesii)*

20b Seed cones have no obvious bracts between scales; leaders droop gracefully.........................
.... western hemlock *(Tsuga heterophylla)*

21a Leaves compound.................................. 22

21b Leaves simple (may be lobed) 37

22a Leaves mostly trifoliate 23

22b Leaves have more than three leaflets............ 25

23a Leaves alternate, strongly scented.....................
.... hopbush *(Ptelea crenulata)*

23b Leaves opposite, not strongly scented. 24

24a Leaf margins irregularly toothed; leaf tips pointed .... box elder *(Acer negundo)*

24b Leaf margins not toothed; leaf tips rounded.......
.... flowering ash *(Fraxinus dipetala)*

25a Leaves palmately compound or nearly so............ 26

25b Leaves pinnately compound 27

26a Leaves palmately compound, arranged along branches of multitrunked trees ....
.... California buckeye *(Aesculus californica)*

26b Leaves deeply palmately slashed, arranged in mop-like clusters at the top of a single trunk ....
.... California fan palm *(Washingtonia filifera)*

27a Leaves opposite.................................. 28

27b Leaves alternate or at the top of a polelike trunk 31

28a Leaves strongly scented; fruits fleshy drupes 30

59b Flowers conspicuous and usually with colorful petals (if in doubt, try going both ways) 70

    60a Fruit an acorn, or a nut enclosed in a spiny burr . . 61

    60b Fruit a capsule or fleshy 66

61a Fruit enclosed in a spiny burr; leaves gold beneath . . . . .
. . . . coast chinquapin (*Chrysolepis chrysophylla*)

61b Fruit an acorn in a scaly or warty cup; mature leaves not gold beneath 62

    62a Leaves bicolored, usually paler underneath 63

    62b Leaves green above and below 64

63a Leaves with saw-toothed margins, more than four inches long . . . . tanbark-oak (*Lithocarpus densiflorus*)

63b Leaves smooth or with spiny-toothed margins, less than four inches long . . . .
. . . goldcup oak or canyon live oak (*Quercus chrysolepis*)

    64a Leaf margins curled under; tiny clumps of hairs at vein junctions on underside of leaves . . . .
. . . . coast live oak (*Quercus agrifolia*)

    64b Leaf margins not curled; no clumps of hairs on underside of leaves 65

65a Leaves shiny yellow green underneath; acorn gradually tapered to its tip . . . .
. . . . interior live oak (*Quercus wislizenii*)

65b Leaves dull olive green underneath; acorn abruptly tapered to its tip . . . .
. . . . Shreve oak (*Quercus parvula* var. *shrevei*)

    66a Male and female flowers in long chainlike catkins
. . . . coast silk-tassel (*Garrya elliptica*)

    66b Flowers not in chainlike catkins 67

67a Leaves gray or gray green 68

67b Leaves green 69

    68a Leaves covered with silvery scales; fruits red . . . .
. . . . Russian-olive (*Elaeagnus angustifolia*)

    68b Leaves not covered with silvery scales; fruits black
. . . . olive (*Olea europea*)

69a Branches drooping; leaves not aromatic . . . .
. . . . maytens (*Maytenus boaria*)

69b Branches upright or spreading; leaves with a distinctive aroma . . . . . . California wax-myrtle *(Myrica californica)*

 70a Flowers with clusters of long, conspicuous, colored stamens; petals small or lacking 70

 70b Flowers with conspicuous petals 74

71a Leaves heavily camphor scented, usually sickle shaped 72

71b Leaves not aromatic, not sickle shaped 73

 72a Leaves bluish green; flowers borne singly . . . . blue gum *(Eucalyptus globulus)*

 72b Leaves greenish; flowers borne in umbels . . . . river red gum *(Eucalyptus camaldulensis)*

73a Leaves (technically, phyllodes) lance shaped, not spine tipped . . . . blackwood *(Acacia melanoxylon)*

73b Phyllodes narrower, sharply spine tipped . . . . Sydney golden wattle *(Acacia longifolia)*

 74a Leaves fragrant when thoroughly crushed 75

 74b Leaves with no special fragrance when crushed 76

75a Leaves strongly bay scented, with entire margins; flowers yellow . . . . California bay *(Umbellularia californica)*

75b Leaves almond scented (crush leaves thoroughly), with serrated margins; flowers white . . . . holly-leaf cherry *(Prunus ilicifolia)*

 76a Flowers white or cream color 77

 76b Flowers some other color 85

77a Flowers urn shaped; bark red or orange 78

77b Flowers disk, bell, or saucer shaped; bark not red or orange 79

 78a Bark red orange; leaves finely serrated . . . . madrone *(Arbutus menziesii)*

 78b Bark red purple; leaves seldom serrated . . . . manzanitas *(Arctostaphylos* spp.)

79a Leaves toothed 80

79b Leaf margins smooth (entire) 81

 80a Leaves glossy; teeth spine tipped . . . . English holly *(Ilex aquifolium)*

80b Leaves dull green; teeth small and not spine tipped
.... toyon *(Heteromeles arbutifolia)*

81a Leaves wavy or with black glands (hold leaf up to the light) 82

81b Leaves neither wavy nor with black glands 83

82a Leaves wavy; flowers orange blossom scented .....
.... Victorian box *(Pittosporum undulatum)*

82b Leaves flat but with black glands; flowers not orange blossom scented........................
.....New Zealand myoporum *(Myoporum laetum)*

83a Flowers cream color, very fragrant....................
.... tobira *(Pittosporum tobira)*

83b Flowers white, not especially fragrant............... 84

84a Thorny side branches; fruits dark purple ....
.... western hawthorn *(Crataegus suksdorfii)*

84b No thorns; fruits orange or red ....
.... cotoneasters *(Cotoneaster spp.)*

85a Flowers blue, purple, or pink 86

85b Flowers yellow or orange 88

86a Flowers a flared trumpet shape, individually large, rose purple or pink........ rosebay rhododendron
.... *(Rhododendron macrophyllum)*

86b Flowers saucer shaped, individually tiny and massed together, blue or purple 87

87a Flowers in long panicles; twigs angled................
.... blue blossom *(Ceanothus thyrsiflorus)*

87b Flowers in rounded clusters; twigs not angled..........
.... jimbrush *(Ceanothus oliganthus var. sorediatus)*

88a Flowers saucer shaped; leaves green............ 89

88b Flowers long and tubular; leaves bluish green .....
.... tree tobacco *(Nicotiana glauca)*

89a Flowers over two inches across; leaves densely felted with irritating hairs.....................................
.... flannel bush *(Fremontodendron californicum)*

89b Flowers no more than half-an-inch across; leaves not felted with irritating hairs............................
.... mountain-mahogany *(Cercocarpus betuloides)*

SELECTED SITES TO OBSERVE BAY AREA TREES

All sites indicated by number on the map are mentioned in the text. Some sites are quite expansive and others are small.

1. Annadel State Park
2. Armstrong Redwoods State Reserve
3. Austin Creek State Park
4. Bear Valley
5. Big Basin State Park
6. Black Diamond Mines Regional Park
7. Bonny Doon
8. Carson Ridge
9. Crystal Springs
10. Harrison Grade
11. Henry Coe State Park
12. Henry Cowell State Park
13. Huckleberry Hill
14. Huckleberry Preserve
15. Inverness Ridge
16. King Mountain
17. Las Trampas Regional Park
18. Mines Road
19. Montara Mountain
20. Morgan Territory
21. Mt. Diablo
22. Mt. Hamilton
23. Mt. St. Helena
24. Mt. Tamalpais
25. Muir Woods
26. Point Reyes National Seashore
27. Pt. Lobos State Reserve
28. Redwood Regional Park
29. Russian River area
30. Salt Point State Park
31. Samuel Taylor State Park
32. San Bruno Mountain
33. Sea Ranch
34. Seventeen-Mile Drive
35. Sugarloaf State Park

Peripheral areas mentioned in the text:

A. Road to Boonville (Highway 128), Mendocino County
B. Hendy Woods, west of Boonville, Mendocino County
C. Van Damme State Park, Mendocino coast

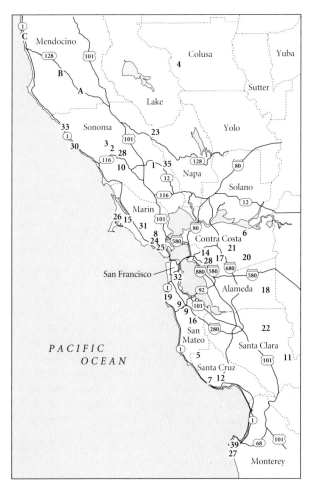

Note: The Santa Lucia Mountains lie south of the Monterey Peninsula and beyond the area encompassed by this map. Likewise, Pinnacles National Monument also lies south of this map.

Aceraceae (Maple Family)

Deciduous shrubs or trees with opposite, often palmately lobed or compound leaves. Flowers are borne in hanging racemes or catkins, sometimes unisexual, sometimes bisexual. Flowers have five separate sepals and petals, or no petals at all, in which case they're wind pollinated. The two-chambered ovary ripens into a doubly-winged samara, which is wind dispersed. *Acer* (maple) is the only genus in this family, except for the tiny Chinese genus *Dipteronia,* and it extends across the Northern Hemisphere. California has four species, the Bay Area two.

BIGLEAF MAPLE   *Acer macrophyllum*
Pls. 1–4

The bigleaf maple usually follows permanent watercourses and so is a typical component of riparian woodlands. It tolerates cool, foggy, coastal conditions as well as the hot summers found inland. With age, the bigleaf maple grows into a stout, often multitrunked tree with a broad canopy that can cover an impressively large area. A single giant bigleaf maple, for example, occurs in the main canyon in Sugarloaf State Park in Sonoma County. The mature bark is deeply fissured into small checkers; the leaves are broad and have three or more deep, palmate lobes (pl. 1). True to its common name, the bigleaf maple has leaves reaching to around six inches across. The leaves start with a lovely reddish tint in early spring and are accompanied by narrow trusses of yellow, bee-pollinated flowers (pl. 2). Leaves turn shades of yellow to orange in fall when there's chill in the night air (pl. 3). Doubly-winged samaras also hang from bigleaf maple branches in fall (pl. 4). Look for the bigleaf maple as a dominant tree in Henry Cowell State Park in the Santa Cruz Mountains, as a common companion in most areas where redwood forests occur, and as far inland as Mitchell Canyon on Mt. Diablo. This tree is

Plate 1. Bigleaf maple. Bigleaf maple's large leaves are deeply palmately lobed and coarsely toothed.

Plate 2. Bigleaf maple. Chains of flowers and new leaves of the bigleaf maple appear midspring.

Above: Plate 3. Bigleaf maple. Bigleaf maple leaves often turn a golden hue in fall.

Right: Plate 4. Bigleaf maple. These doubly-winged bigleaf maple samaras

found beyond our region, northward, southward, and into the middle elevations of the Sierra Nevada.

BOX ELDER *Acer negundo* var. *californicum*
Pls. 5–7

Although the box elder is found in every state of the Union, California's version is different enough to merit this varietal name. Whereas the bigleaf maple *(A. macrophyllum)* is to be expected in many riparian areas, the box elder has a much more scattered distribution, appearing only on the broadest, best-developed floodplains of major streams and rivers. Two excellent examples are along the Russian River in Sonoma County and by the San Lorenzo River in Henry Cowell State Park in the Santa Cruz Mountains. Although the overall form of the box elder is similar to that of the larger maples—a broad, rounded canopy with multiple trunks—it is immediately distinguished by its compound leaves. Each leaf consists

Plate 5. Box elder. Box elder leaves are light green and compound, with jagged teeth.

Left: Plate 6. Box elder. The gray brown bark of the box elder is irregularly checkered on mature trees.

Above: Plate 7. Box elder. Mature, brown box elder samaras may linger on branches after the leaves have dropped.

of three to five pinnately arranged leaflets—that is, the terminal leaflet is on a stalk that extends beyond the level of the lateral leaflets (pl. 5). Box elder leaves seldom display the color changes typical of the bigleaf maple in spring or fall. Another special aspect of the box elder is that the trees are dioecious, that is, each tree is either male or female. In early spring, male box elders are festooned with long slender catkins of pinkish, stamen-bearing flowers, whereas the female's flowers are most conspicuous in fall when their pendant racemes of doubly-winged samaras have ripened (pl. 7). This fast-growing tree is sometimes cultivated in a varietal form that has pink-variegated leaves.

Anacardiaceae (Sumac Family)

Trees or shrubs that are either deciduous or evergreen. The leaves and stems often have resin-containing ducts that give a distinctive odor; sometimes the sap is corrosive or raises blisters, as with poison-oak *(Toxicodendron diversilobum)*. Leaf patterns vary tremendously, although many species have trifoliate or pinnately compound leaves. Tiny, whitish, greenish, or pinkish flowers are borne in dense clusters—either in racemes or panicles. Each flower consists of five separate sepals and petals together with five or 10 stamens and a single pistil with a superior ovary. The fruits are fleshy, one-seeded drupes, sometimes with large, nutlike seeds. Several interesting edible plants belong to this family, including the mango *(Mangifera indica)*, pistachio *(Pistacia vera)*, and cashew *(Anacardium excelsum)*. California is home to a few native shrub genera that include species such as poison-oak and sugar bush *(Rhus ovata)*, but the only tree is the Chilean-pepper tree *(Schinus molle)*, which is sometimes naturalized.

CHILEAN-PEPPER TREE · *Schinus molle*
Pls. 8–10

NOT NATIVE

The Chilean-pepper tree is one of the most widely cultivated trees in coastal California, where it thrives in droughty gardens. So characteristic is it in our landscapes that it is often incorrectly known as the California-pepper tree, implying a connection to our flora that doesn't exist. The Chilean-pepper tree is easy to identify by its gnarled trunk (pl. 8); its rounded crown with long gracefully drooping branches (they may occasionally touch the ground) (pl. 9); the evergreen, pinnately compound leaves composed of many narrow leaflets (pl. 10); the panicles of tiny, whitish flowers; and the bright red, peppercorn-like fruits that follow. These fruits give the tree its common name and create a mess on sidewalks. The true source of our condiment pepper is a tropical vine *(Piper nigrum)* belonging to the pepper family (Piperaceae). The Chilean-pepper tree is often encountered in canyons and abandoned fields, especially in southern California, but it may also be seen inland in the Bay Area, its seeds having been carried by the birds that feed on the fruits.

Plate 8. Chilean-pepper tree. Chilean-pepper tree bark is often knotted.

Plate 9. Chilean-pepper tree. The outline of the Chilean-pepper tree is broad with many gracefully drooping branches.

Plate 10. Chilean-pepper tree. The drooping, feathery, pinnately compound leaves of the Chilean-pepper tree are quickly recognizable.

Aquifoliaceae (Holly Family)

Usually dioecious shrubs or trees with simple, alternate, toothed, lobed, or spiny leaves. Flowers are small and white or greenish and borne in small clusters in leaf axils. Each flower has four sepals, four petals, four stamens, and a pistil with a superior four-chambered ovary. The fruit is a four-seeded, berrylike drupe. This modest-sized family is widely distributed in forested areas, both in the tropics and the temperate zone. It is best known for the ornamental English holly *(Ilex aquifolium)* and the yerba maté, a South American species whose leaves produce a stimulating beverage. None of the family is native to California, but the English holly occasionally grows wild in the Bay Area.

ENGLISH HOLLY   *Ilex aquifolium*
Pls. 11, 12   NOT NATIVE

Because it's the quintessential emblem of Christmas, the English holly is widely cultivated in the Bay Area. The bright red berries are attractive to birds and so are dispersed by them to natural areas where they sometimes take root and grow. The English Holly succeeds especially well in coastal forests and is to be expected on Inverness Ridge in Pt. Reyes National Seashore and in the Berkeley and Oakland hills. Although many people think of it as a shrub, it can grow into a 20- to 30-foot tree. The English Holly is immediately identified by its thick, glossy, wavy, dark green leaves edged with viciously spiny teeth (pl. 11). Small, white blossoms appear in spring (pl. 12), with male plants bearing staminate flowers, and female plants bearing pistillate flowers. Only the female plants produce the vivid red berries that are such a familiar sight in late fall and early winter.

Plate 11. English holly. Here you can see the spiny teeth and slightly twisted aspect of typical English holly leaves.

Plate 12. English holly. English Holly leaves are a glossy, deep green. The white flowers are borne in tight clusters in the center of the leaves.

Arecaceae (Palm Family)

Shrubs to tall trees, usually with a single, uniform, un-branched trunk topped by a mop of large, palmately lobed to pinnately compound leaves. Flowers are borne in complex racemes or panicles, often among the leaves or just below them. Palm flowers are usually white or cream in color and composed of three sepals and petals that look alike, six sta-mens (sometimes more), and a usually superior ovary that ripens into a fleshy drupe. Palm fruits range in size from pealike to basketball-sized. Palms are important food plants in the tropics (most palms are subtropical or tropical), the most prominent being various oil palms (palm oil is widely used in foodstuffs and for other applications), coconut palms, and date palms. Although this is a large, important family, only the California fan palm *(Washingtonia filifera)* is native to southern California, but several species are cultivated in the Bay Area. Occasionally, the Canary Island date palm *(Phoenix canariensis)* is naturalized in coastal areas, the seeds having been carried there by birds eating the fruits.

CANARY ISLAND DATE PALM   *Phoenix canariensis*

Pls. 13, 14   NOT NATIVE

Closely related to the commercial date palm *(P. dactylifera)*, the Canary Island date palm is restricted to the mild-climated Canary Islands off the coast of North Africa. Because it is hardier than most other palms that are planted in California, you may see this palm as far north as Eureka. In the Bay Area, birds sometimes carry the seeds to mild coastal sites after fin-ishing a meal of the sweet, datelike fruits. The Canary Island date palm has 15- to 20-foot-long, pinnately compound fronds; drooping panicles of white flowers; and small, dark brown, datelike fruits. Of the many feather-leaved palms (palms with pinnately compound leaves), the phoenixes are distinguished by leaflets near the base of the frond that have

Plate 13. Canary Island date palm. The Canary Island date palm is typified by segments at the base of each leaf that are modified into stout spines.

Plate 14. Canary Island date palm. The Canary Island date palm has a trunk that bears one oversized, moplike crown of gracefully drooping, feathery, pinnately compound leaves.

been modified into dagger-sharp spines (pl. 13). The bases of the young trees resemble large pineapples, especially where the oldest leaves have been trimmed away to expose the orange brown trunk. Mature trees bear a massive trunk with a moplike crown of gracefully drooping leaves (pl. 14).

Betulaceae (Birch Family)

Deciduous shrubs or trees with simple, alternate leaves that are often doubly serrate, having both fine and coarser sawlike teeth. The trees are monoecious—male and female flowers are borne on the same plant—and depend on wind pollination. Male flowers are carried in dangling, chainlike catkins and lack colorful petals; female flowers are either borne in shorter upright catkins or are in small clusters along the twigs. Male flowers consist of bracts and long stamens; female flowers consist of bracts and an inferior ovary with three feathery stigmas. The female catkin ripens into a small conelike structure filled with tiny, one-seeded achenes, or, in the case of the California hazelnut *(Corylus cornuta* var. *californica),* a hard-shelled, edible nut surrounded by fuzzy bracts. This small family is adapted to moist places and is mostly limited to riparian areas in California. California has three genera: *Betula* (birch), *Alnus* (alder), and *Corylus* (hazelnut), but only alders and the California hazelnut *(C. cornuta* var. *californica)* occur in the Bay Area.

WHITE ALDER   *Alnus rhombifolia*

Pls. 15–17

Alders are slender, fast-growing trees that follow permanent watercourses. There are far more traits shared among the different alder species than there are differences. Alders have nitrogen-fixing nodules on their roots that allow them to grow in soils low in nutrients, and they are able to invade newly disturbed sites. Alder bark is usually smooth and gray (pl. 15);

Plate 15. White alder. Copses of white alders in winter display the pale whitish bark.

leaves are ovate and doubly serrate (pl. 16), with intricate, interjoined veins. Although the leaves may turn pale yellow in fall, they're seldom noted for their color. Long, narrow, dangling male catkins appear in late winter or earliest spring before leaves unfurl from buds. Smaller, stubby, upright, female catkins open a bit later than the males to prevent self-pollination. In fall, female catkins ripen into conelike struc-

Right: Plate 16. White alder. White alder leaves are characterized by a doubly-serrated edge and a complex and characteristic vein pattern.

Below: Plate 17. White alder. White alder crowns are relatively narrow and rounded. These trees are just leafing out in spring in the riparian woodland of Mitchell Canyon on Mt. Diablo.

tures that look like miniature redwood seed cones. Each "cone" contains many tiny, one-seeded, winged achenes for broad and efficient wind dispersal to new sites. The white alder typically replaces the red alder *(A. rubra)* in summer-hot inland areas. The two species are closely similar, but white alder leaves are flat and lack rusty hairs underneath. Look for the white alder in Sugarloaf State Park, along Mitchell Canyon on Mt. Diablo (pl. 17), in Redwood Regional Park in the Oakland Hills, and along streams in the Mt. Hamilton area.

RED ALDER   *Alnus rubra*

Pls. 18–20

As noted above, the red alder closely resembles its sister species, the white alder *(A. rhombifolia),* including dangling male catkins in late winter (pl. 18) and conelike seed catkins in fall (pl. 19). The red alder typifies cool coastal canyon bot-

Plate 18. Red alder. The red alder dangles long, slender male catkins before new leaves appear.

Plate 19. Red alder. The ripe seed catkins of the red alder resemble tiny, redwood seed cones.

Plate 20. Red alder. This dense stand of red alders follows a watercourse at Sea Ranch. Notice the leafless winter condition. Coast redwoods *(Sequoia sempervirens)* stand behind these red alders on higher ground.

toms and is often prominent along stream corridors in redwood country (pl. 20). Further north, where rainfall is more plentiful, it moves in as a pioneering tree when redwoods have been logged out. Beneath the alder cover, of course, young redwood saplings and stump sprouts flourish and will later shade out the alders to reestablish the original forest. The red alder is so named because its roots have a decidedly red bark, which the Indians used as a red dye to create designs in their baskets. Although red alder trunks are closely similar to those of the white alder, they are often obscured by a complex tapestry of epiphytic plants—leafy liverworts, polypody ferns, mosses, and lichens—that thrive in the humid air of coastal locales. The best traits for distinguishing between the two alders are the leaf details: red alder leaf margins turn under, and the undersides of the leaves have scattered rusty hairs. In most other respects, the two species are closely alike. Visit Armstrong Redwoods State Reserve, Muir Woods, or the coastal canyons in Pt. Reyes National Seashore to see splendid copses of red alders.

CALIFORNIA HAZELNUT
Corylus cornuta var. *californica*

Pls. 21–23

Closely related to the commercial filbert, the California hazelnut is at once identified as belonging to the same family as alders by its ovate, double-serrate leaves (pl. 21), and its long, narrow male catkins that open in late winter (pl. 22). But there

Plate 21. California hazelnut. California hazelnut leaves are soft and furry but otherwise resemble those of the related alders. The young nut here is covered by fuzzy, beaked bracts.

Plate 22. California hazelnut. The slender male catkins of the California hazelnut open to shed their pollen while branches are still devoid of leaves.

Plate 23. California hazelnut. This close view shows the tiny female flowers with their extended, red stigmas designed to catch wind-borne pollen.

the similarities end, for the California hazelnut is more a multibranched large shrub growing up to perhaps 15 or 20 feet tall, with soft, downy leaves (some wits refer to these as "Indian" toilet paper); small clusters of female flowers topped with feathery, dark red stigmas (pl. 23); and large edible nuts. Each nut, which ripens from late summer to early fall, is guarded by a long, beaked pair of bracts whose stiff, prickly hairs are unpleasant to handle (pl. 21). Although the nuts are delicious, the California hazelnut seldom bears a large crop in our area, and birds and squirrels usually find the good nuts before they're discovered by humans. Shortly after the nuts have been harvested, the leaves turn to shades of faded gold before they drop. Look for the California hazelnut along the edges of closed-cone pine or redwood forests, often near stream courses. Its intricate branches are arrayed in horizontal tiers —another good identifying feature when the tree is leafless in winter. The California hazelnut is also seen in the yellow pine (*Pinus ponderosa*) belt of the Sierra and in northern mountains.

Caprifoliaceae (Honeysuckle Family)

Mostly deciduous shrubs and small trees with opposite, sometimes pinnately compound leaves. Flower size and design vary considerably; often, small flowers are arranged in complex cymes, panicles, or in other arrangements. Flowers have five minute sepals (often difficult to see), five partly fused petals, five stamens, and an inferior ovary with two chambers. Fruits are most often fleshy, drupelike berries attractive to birds and mammals. This Northern Hemisphere family includes several shrubs in California and small trees—the elderberries (*Sambucus* spp.)—that enter our area. Several members of the family, including viburnums and honeysuckles, are widely grown in gardens.

BLUE ELDERBERRY   *Sambucus mexicana*
Pls. 24, 25

The blue elderberry grows as either a large multibranched shrub or a small tree up to 20 feet tall. Single-trunked specimens develop fluted bark, but there are often multiple trunks. New shoots from the base grow tall and straight and were used by the Indians as clappers, gambling sticks, and flutes, made by hollowing out the soft central pith. All elderberries are noted for their large, pinnately compound leaves, whose odor elicits various characterizations ranging from "stinky" to "pleasant." The blue elderberry is readily identified by its flat-topped cymes of fragrant, cream-colored, late-spring flowers (pl. 24), attractive to many different pollinators and

Plate 24. Blue elderberry. The pinnately compound leaves and flat-topped clusters of tiny, whitish flowers clearly identify the blue elderberry.

Plate 25. Blue elderberry. By midsummer, the blue elderberry's flowers are replaced by clusters of tiny, pale blue, edible fruits.

filled with nectar. Dull, bluish berries ripen (pl. 25) in summer and are eaten raw, cooked, made into jam, or fermented into wine. Among the Indians, the blue elderberry was also noted for its medicinal uses. Look for the blue elderberry in inland valleys, often close to watercourses, where it grows rapidly to maturity. It occurs in many parts of Marin County, the Mt. Diablo region, along Mines Road, on the Peninsula, and throughout Sonoma County.

RED ELDERBERRY *Sambucus racemosa*
Pls. 26, 27

The red elderberry prefers foggy, shaded coastal canyons where it luxuriates in the company of alders, maples, and redwoods. It bears a close resemblance to the blue elderberry

Plate 26. Red elderberry. The red elderberry has coarse, pinnately compound leaves and pyramidal clusters of whitish flowers.

(S. mexicana), but the leaves are even larger, and the snowy white flowers appear in early to midspring, arranged in pyramidal clusters (pl. 26). In summer, the red elderberry is made even more striking by its vivid red berries (pl. 27), which attract many kinds of birds, but these berries often

Plate 27. Red elderberry. The bright red fruits of the red elderberry are tempting to birds but may make humans sick.

make humans sick. In other respects—straight, soft-pithed twigs and medicinal leaves—the red elderberry was used in much the same way as the blue elderberry. An interesting use is as an antidote to nettle stings: the crushed leaves are applied to the skin. A variety of red elderberry occurs in the high mountains, but it is a low shrub that favors the edges of rocky meadows; botanically, it is known as the dwarf elderberry *(S. racemosa* var. *microbotrys).*

Celastraceae (Staff-tree Family)

Shrubs or trees with often opposite, simple leaves. The flowers are arranged in various ways and have a shallow, cup-shaped hypanthium with four to five sepals and petals, four to five stamens attached to a disk, and a superior two- to five-chambered ovary partially embedded in the disk. Fruits vary widely but are often fleshy or capsules with seeds that bear fleshy appendages (arils). This moderate-sized family has a worldwide distribution but is particularly abundant in eastern Asia. A few ornamentals are cultivated, such as the burning-bush *(Euonymus* spp.) and the maytens *(Maytenus boaria).* Even though the maytens is not listed in *The Jepson Manual* (Hickman 1993), it is considered to be a local and invasive pest in parts of the Bay Area.

MAYTENS   ***Maytenus boaria***
Pls. 28, 29   NOT NATIVE

Not many of the trees we grow come from Chile, but the maytens is one. Actually, central Chile has a Mediterranean climate strikingly similar to our own, but most of its ornamental plants are poorly known here. At first glance, you might mistake the maytens for a weeping willow *(Salix babylonica)*—the broad rounded crown and drooping branches that nearly touch the ground are strikingly similar (pl. 28). But maytens is an evergreen tree with very small, angled

Plate 28. Maytens. The maytens has a broad crown with drooping branches much like that of the weeping willow *(Salix babylonica),* but its leaves are firm and evergreen.

leaves seldom exceeding two inches in length (pl. 29), and the leaves bear no stipules the way most willow leaves do. Another striking feature is the vigorous suckers that develop from the roots. Finally, the rather warty bark is different from that of any willow. The tiny, greenish flowers are seldom noticed, but sometimes the female trees ripen fruits. Each pea-sized fruit opens to display a few seeds covered with a fleshy, red aril at-

Plate 29. Maytens. Maytens leaves are narrowly lance shaped, and the minute flowers appear in clusters among the leaves.

tractive to birds. Maytens has recently come to the attention of native plant enthusiasts, who have found it spreading on its own in some of the East Bay hills. Its strong, vigorous suckers may be partly responsible, but probably the original trees came from seeds carried to wild places by birds feeding on the fruits in someone's garden.

Cornaceae (Dogwood Family)

Mostly deciduous shrubs or small trees with opposite, ovate to elliptical leaves that are smooth margined, are not toothed, and have a distinctive arcuate-pinnate vein pattern. New twigs are often reddish. Tiny. greenish or whitish flowers are borne in flat-topped cymes or in dense, buttonlike heads. The flowers are sometimes surrounded by showy, petal-like bracts and consist of four minute sepals, four separate petals, four stamens, and an inferior ovary that ripens into a fleshy, berry-like drupe. The family has a strong northern distribution and usually occurs in moist, shaded places. Our only native genus is *Cornus* (dogwood), and most species are multitwigged shrubs. The Pacific flowering dogwood *(C. nuttallii)* is a small tree included here.

PACIFIC FLOWERING DOGWOOD *Cornus nuttallii*
Pls. 30–33

When easterners see our flowering dogwood, they're reminded of *C. florida* from the southeastern states. This dogwood is more easily cultivated than our own native and often appears as an ornamental in Bay Area gardens. The Pacific flowering dogwood grows tall and narrow with many tiers of horizontally trending branches (pl. 30). New leaves appear in early to midspring well before the blossoms open, although the flower buds are already present at that time. The real show comes in late April to May, when dozens of white saucers dec-

Plate 30. Pacific flowering dogwood. The Pacific flowering dogwood tree has a narrow shape, a bright green color, and many horizontally tiered branches.

Plate 31. Pacific flowering dogwood. In late April to early May the many horizontal branches of the Pacific flowering dogwood are decorated with saucer-sized clusters of beautiful, white-bracted flowers.

orate the branches (pl. 31). Each saucer actually consists of six large, white, petal-like bracts surrounding a mound of small, green, starlike flowers. A second show occurs in late summer when red orange, ball-shaped fruit clusters ripen (pl. 32). The final show arrives in mid- to late fall when leaves glow in shades of orange, rose purple, or red before falling (pl. 33). The Pacific flowering dogwood is rare in much of the Bay Area; it is very localized, for example, in the Santa Cruz Mountains and is entirely missing from the East Bay. But if you travel to Mt. St. Helena in spring, you'll be rewarded with beautiful displays of blossoms among the ponderosa pines *(Pinus ponderosa)*, sugar pines *(P. lambertiana)*, and California black oaks *(Quercus kelloggii)*. Ida Clayton Road is a particularly good place to see these dogwoods.

Left: Plate 32. Pacific flowering dogwood. The multiple fruits of the Pacific flowering dogwood turn red orange by summer's end.

Below: Plate 33. Pacific flowering dogwood. Pacific flowering dogwood leaves turn luminous shades of red in fall.

Cupressaceae (Cypress Family)

Coniferous trees (occasionally shrubs) with whorls of needle-like juvenile leaves and pairs or whorls of tiny, scalelike adult leaves. Branches may have a whiplike or three-dimensional quality, or they may appear flattened, resembling fern fronds. The pollen cones are minute and inconspicuous; the seed cones are egg to globe shaped and have woody scales, or become fleshy and resemble berries. This prominent conifer family is widely distributed throughout the Southern and Northern hemispheres; California has four genera and several species. Some, like the western red cedar *(Thuja plicata)* are noted for their fragrant and durable wood. In the Bay Area, we have the fleshy-coned junipers *(Juniperus* spp.); the flat-branched, red-barked incense-cedar *(Calocedrus decurrens);* and several species of cypress *(Cupressus* spp.), with tightly closed, woody seed cones. Cypresses grow in unusual, nutrient-poor soils; several species are very rare.

INCENSE-CEDAR   *Calocedrus decurrens*

Pls. 34–36

The incense-cedar is a typical component of the mixed-conifer belt throughout most of California's mountains, especially on the west side. It enters the Bay Area in sparing fashion, appearing occasionally on serpentinite soils in Mendocino and Napa counties. The incense-cedar is a handsome tree that is easily told by many different traits: cinnamon-colored bark in long, fibrous strips; vivid, yellow green foliage arrayed in flattened, fernlike sprays (pl. 34) (vertically ori-

Plate 34. Incense-cedar. The incense-cedar is typified by flattened branches that have a fernlike appearance. These branches bear dozens of minute pollen cones in fall.

Plate 35. Incense-cedar. Before they fully open, these hanging seed cones of the incense-cedar look like comical duck bills.

ented in sunny positions; horizontally in shaded ones); and brown, spindle-shaped seed cones that open by their two outer scales into silhouettes of birds in flight (pl. 35). In youth, the trees are a picture-perfect conical shape (pl. 36), but in old age the crowns become rounded and craggy. Few

Plate 36. Incense-cedar. The incense-cedar is noted for its trim, conical shape when young.

trees in the Bay Area have attained the 100-foot-high stature that is commonly seen in the northern mountains or in the Sierra. Not a true cedar at all, the incense-cedar is aptly named, at least, for its fragrant foliage and wood. Even the genus name suggests a cedarlike nature, for it translates as *beautiful cedar*. Look for the incense-cedar along Hwy. 128 between Cloverdale and Boonville in southern Mendocino County, and along Hwy. 29 north of Mt. St. Helena on the way to Middletown.

SANTA CRUZ CYPRESS *Cupressus abramsiana*
Pl. 37

All of California's cypresses—minus those species that are often described as belonging to the genus *Chamaecyparis*—have whiplike branches with a three-dimensional arrangement of tiny, fragrant, scalelike leaves (pl. 37). All are noted for their minute pollen cones borne in great quantity, usually in fall, and for their egg- to globe-shaped, woody seed cones that stay tightly fastened to the trees for life, and which generally open only after fire. The shield-shaped scales of these seed cones are arrayed in whorls; the center of each bears a raised, nipplelike umbo. Cypresses are survivors on nutrient-poor, often arid and rocky soils, where other trees can't compete. Some, like the Santa Cruz cypress, have an extremely restricted range. In fact, so rare is the Santa Cruz cypress, that it occurs in only four localities, all within the southern part of the Santa Cruz Mountains. The best stand, with the greatest number of trees, occurs on Martin Road just a mile or so northeast of Bonny Doon. There it grows in company with the ponderosa pine *(Pinus ponderosa)*, coast live oak *(Quercus agrifolia)*, and knobcone pine *(P. attenuata)* on soils derived from ancient sand dunes. In addition to this odd mixture of trees, these sand dunes support a bizarre maritime chaparral that includes the rare and beautiful silver-leaved manzanita *(Arctostaphylos silvicola)*.

Plate 37. Santa Cruz cypress. The rare Santa Cruz cypress sends out long, whiplike branches when it's young. This tree is part of a large population near the town of Bonny Doon in the Santa Cruz Mountains.

GOWEN CYPRESS

Pls. 38–40

Cupressus goveniana
subsp. _goveniana_

The Gowen cypress is even rarer and more exacting in its habitat than the Santa Cruz cypress *(C. abramsiana)*. In its typical form—a shrubby, mounded tree that grows up to 15 or 20 feet high (pl. 39)—it occurs in only two places: at the eastern end of Pt. Lobos State Reserve south of Carmel, and on the crown of Huckleberry Hill near Seventeen-Mile Drive. Both sites have ancient, highly leached, white podzols. A vari-

Plate 38. Gowen cypress. The seed cones of the Gowen cypress are often densely clustered and nearly globular in shape. Each cone remains tightly closed for many years.

ant known as the pygmy cypress *(C. goveniana* subsp. *pygmaea)* is found nearly 100 miles to the north on ancient marine terraces with even shallower podzols. These shallow soils are poorly drained because they're underlain by a cementlike hardpan, and the vegetation that survives there is so stunted it's called the pygmy forest. The pygmy cypresses that grow on these podzols reach to less than 10 feet tall yet bear normal-sized seed cones (pl. 40). When these same cypresses grow on more normal soils, as they sometimes do among bishop pines

Plate 39. Gowen cypress. The Gowen cypress is characterized by a bright green color and a rather rounded crown. The tree seldom exceeds 15 or 20 feet in height.

Plate 40. Pygmy cypress. The hand and arm in this photo give a sense of how small these 100-year-old pygmy cypress trees are in their natural habitat at Van Damme State Park on the Mendocino coast.

(Pinus muricata) and redwoods, they soar to over 100 feet high. The Gowen form of this cypress has a dense, bushy shape, whereas the pygmy form is usually slender and narrow in habit, but otherwise the two varieties are closely similar. To see the pygmy cypress in its stunted form, drive to the upper end of Van Damme State Park on the Mendocino coast; walk up the trail from Jughandle Botanical Reserve near Caspar; or climb the trail at Salt Point State Park along the northern Sonoma coast.

MACNAB CYPRESS *Cupressus macnabiana*

Pls. 41–43

Whereas the Santa Cruz, Gowen, and Monterey cypresses *(C. abramsiana, C. goveniana* subsp. *goveniana,* and *C. macrocarpa)* are coastal, the Macnab and Sargent *(C. sargentiana)* cypresses grow inland where the summers are hot and bone dry. Both are restricted to serpentinite soils, and both are drought adapted, but the Macnab cypress is most consistently associated with the driest, hottest sites. The Macnab cypress has an arclike distribution that runs from the inner North

Plate 41. Macnab cypress. These branches of the Macnab cypress are loaded with minute, yellow pollen cones that shed their pollen to the winds.

Plate 42. Macnab cypress. The Macnab cypress often grows broader than tall, as seen here.

Plate 43. Macnab cypress. This close-up view of Macnab cypress twigs shows how tiny each scalelike leaf is. A white powder covers the leaves and imparts to them a strong fragrance.

Coast Ranges of Napa and Lake counties over to the northern Sierra foothills. The Macnab cypress is consistently a low, multitrunked tree with many horizontally trending branches and so is often broader than tall (pl. 42). It is easily recognized by its bluish hue, which is caused by tiny glands on the backs of the leaves that excrete a highly aromatic, resinous powder (pl. 43). This resin protects the leaves because of its unpleasant taste, and it reflects the harsh summer sun. Look for the

Macnab cypress on Walker Ridge along the west side of Bear Valley in eastern Lake County, and on Ida Clayton Road on the northeast side of Mt. St. Helena.

MONTEREY CYPRESS   *Cupressus macrocarpa*

Pls. 44, 45

The Monterey cypress is quite possibly the best known of the world's cypresses and is symbolic of the Monterey Peninsula, to which it is naturally restricted. In fact, this rare cypress is found in just two groves: along Seventeen-Mile Drive in Carmel and to the south in Pt. Lobos State Reserve (pl. 44). The Monterey cypress is widely planted in cool coastal areas,

Plate 44. Monterey cypress. Point Lobos State Reserve is a wonderful place to see Monterey cypresses in their natural habitat. They cling precariously to rocky bluffs by the sea.

Plate 45. Monterey cypress. Monterey cypress branches are deep green and mostly horizontally arrayed.

not only along our own coast but in such far-flung places as Australia and New Zealand. It creates a fast-growing windbreak and was an important element in taming the winds to make Golden Gate Park feasible. It is also tolerant of salt spray and poor, sandy soils. Despite these appealing qualities, the Monterey cypress may reseed aggressively on its own and claim territories that would otherwise support more appropriate locally native vegetation. The Monterey cypress is typified by a beautifully wind-sculpted crown with many horizontally splayed branches, deep green foliage (pl. 45), and seed cones to three inches long (*macrocarpa* means large cone). Short-lived in its natural habit, many of the Monterey cypresses on the Monterey Peninsula are showing signs of aging and are clearly in decline, for they need to be renewed by fire.

SARGENT CYPRESS   *Cupressus sargentia*

Pls. 46, 47

The Sargent cypress is the most widespread of California's cypresses, occurring from San Luis Obispo County in the south to Lake County in the north. Confined to serpentinite soils and barrens, many large stands exclude practically all other woody plant life. Unlike the Monterey cypress *(C. macrocarpa),* the Sargent cypress often has a crown taller than broad, and the branches don't all extend horizontally (pl. 46). Although it sometimes grows within sight of the Macnab cypress *(C. macnabiana)* in the north, the two are immediately distinguished by color and shape: the Macnab cypress is a squat, shrubby tree with whitish green foliage; the Sargent cy-

Plate 46. Sargent cypress. This photo shows the dense, bushy, dark green branches of Sargent cypress trees. Here they're growing with gray pines *(Pinus sabiniana).*

press is taller and a much richer green without the white resins that accumulate on Macnab cypress leaves (pl. 47). Look for the Sargent cypress east of Occidental on Harrison Grade in Sonoma County, and on Ida Clayton Road by Mt. St. Helena. To see a truly large stand of dwarfed trees, visit the Carson Ridge area of Marin County, where several square miles are covered with nothing but this cypress.

Plate 47. Sargent cypress. Sargent cypress branches show tiny, scale-like leaves that are dark green rather than the whitish color of the Macnab cypress. Both cypresses may grow in the same area of the inner parts of Lake County.

CALIFORNIA JUNIPER   *Juniperus californica*
Pls. 48–50

Many tree books fail to mention the California juniper because it so often grows as a large shrub (pl. 48), whereas the Sierra and Utah junipers *(J. occidentalis* and *J. osteosperma)* are full-fledged trees. Nonetheless, it shouldn't be summarily dismissed, for it often develops one or a few main trunks in

Plate 48. California juniper. The California juniper often appears like a rounded, overgrown shrub. This one is growing in the arid interior foothills of the inner South Coast Ranges.

tree fashion, even if it seldom grows more than 20 feet tall. The California juniper also has the broadest distribution of any juniper in California, extending from the southern desert mountains into the southern Sierra foothills and marching along the hottest inland foothills of the Coast Ranges all the way north to Tehama County near Redding. In overall appearance, the California juniper is typical of other junipers: stringy strips of brownish bark; three-dimensional twigs carrying tiny, fragrant, deep green, scalelike leaves (pl. 49); and fleshy, pale purple seed cones commonly called berries (pl. 50). The California juniper is usually dioecious, so separate trees bear either tiny pollen cones or berries. As with all junipers, the berries carry a scent reminiscent of gin, which is flavored by the berries of a subtropical juniper. The California juniper is abundant on Mt. Diablo, where place names such as Juniper Camp reveal its presence. Look for it also along Mines Road.

Plate 49. California juniper. Like the cypresses, the California juniper has twigs covered with minute, scalelike leaves.

Plate 50. California juniper. Although California juniper branches look much like those of cypresses, they're easily recognized by the pale bluish, berrylike seed cones.

Elaeagnaceae (Oleaster Family)

Usually deciduous shrubs or trees with simple, alternate or opposite, entire leaves covered with silvery scales. Small flowers are borne in umbel-like clusters with four petal-like sepals topping a fleshy hypanthium, and four stamens directly attached to the hypanthium. Single-chambered ovaries are surrounded by the hypanthium so that they look inferior, and the whole thing—hypanthium and ovary—ripens into a fleshy, drupelike fruit. This small family occurs across the Northern Hemisphere and south into Australia. It is noted for ornamental shrubs in the genus *Elaeagnus* and for the native buffalo berry *(Shepherdia argentea)* from the high desert. The so-called Russian-olive *(Elaeagnus angustifolia)* is described below because it is occasionally encountered growing "wild" in the hotter parts of the Bay Area.

RUSSIAN-OLIVE   *Elaeagnus angustifolia*

Pls. 51–52   NOT NATIVE

The Russian-olive—a tree of Asiatic origin—is often planted in hot, arid places, especially where wind or salt in the soil is a problem. Because the inland portion of the Bay Area sometimes has these conditions, you may see the Russian-olive there as a small street tree or occasionally naturalized in the inner Coast Ranges. Not an olive at all, this deciduous tree has narrow, grayish, olivelike leaves (pl. 51) and produces olive-shaped fruits. You can quickly tell the difference between a Russian-olive and the real thing by the Russian-olive's shaggy brown bark, its angled twigs, and its fruits—up close you can see the red fruits are prettily covered with silver scales, a hallmark of the oleaster family. The small clusters of yellow green flowers are fragrant and appear in spring. The Russian-olive bears a broadly rounded crown (pl. 52).

Plate 51. Russian-olive. Russian-olive leaves are shaped much like those of willows (*Salix* spp.) and the edible olive, but they're densely covered with silvery scales.

Plate 52. Russian-olive. The crown of the Russian-olive is broadly rounded, and the smaller twigs droop.

Ericaceae (Heather Family)

Woody ground covers, shrubs, or trees—often evergreen. Bark patterns and colors are often diagnostic for certain groups in this family (for example, smooth, red purple bark on manzanitas). The simple leaves are often leathery, sometimes toothed, sometimes entire. Flowers may be tubular, bell shaped, or urn shaped or have a flared, trumpet shape. (Species with urn-shaped flowers are particularly common in California.) Flowers have four or five small sepals, four or five partly joined petals, four to 10 stamens, and a superior (occasionally inferior) ovary that ripens into a several-chambered capsule with tiny seeds or into a fleshy, often edible berry. The most distinctive feature of the family is the two holes found at the ends of the stamens through which the pollen is shed (use a good hand lens); the stamens of most other flowers open by lengthwise slits. This large family occurs in the Northern Hemisphere and on mountains in the tropics and is well represented in California. Many species are adapted to cool conditions and acid soils. Few of our natives are trees, but the Bay Area is home to madrone *(Arbutus menziesii)*, noted for its peeling, red orange bark and finely serrated leaves; a few treelike manzanitas *(Arctostaphylos* spp.), identified by their red purple bark and entire, usually vertically oriented leaves; and rosebay rhododendron *(Rhododendron macrophyllum)*, noted for its large rose purple, trumpet-shaped flowers.

MADRONE   *Arbutus menziesii*

Pls. 53–56

The madrone is the largest member of the heather clan in California and is a stout, often-multitrunked tree that reaches over 100 feet high. Usually a component of mixed-evergreen forests, common companions include tanbark-oak *(Lithocarpus densiflorus)*, Douglas-fir *(Pseudotsuga menziesii)*, California bay *(Umbellularia californica)*, and coast live oak *(Quercus agrifolia*). The madrone is immediately identified by its bark: the newest bark is as smooth as a baby's skin and greenish to pale tan. On a hot summer day, the bark is refreshingly cool to the touch because of the water being carried upward just beneath. The bark quickly turns orange brown and, with age, flakes off in narrow curls (pl. 53). For this reason, the base of an old madrone has a very different appearance from the upper limbs and branches. The madrone is also noted for its uncommonly handsome, elliptical, evergreen leaves with finely serrated edges (run your finger along the leaf margin) and silvery backs. The leaves are usually obliquely oriented in summer so that their pale undersides can reflect away the heat from the summer sun. Panicles of sweetly fragrant, jug-shaped, white blossoms appear from early to midspring (pl. 54) and are a strong draw to bees. By November, the fruits

Plate 53. Madrone. The pale orange bark of the madrone peels, curls, and darkens as it ages.

Plate 54. Madrone. The white, urn-shaped, fragrant blossoms of the madrone are arrayed in open, panicle-type sprays.

Plate 55. Madrone. The madrone has narrowly oval leaves and clusters of warty, red orange berries.

Plate 56. Madrone. Madrone trees often top the crowns of rocky hills in Marin County, especially near Carson Ridge.

have ripened into warty, red orange berries attractive to birds (pl. 55). To humans, madrone berries are dry and tasteless. Madrones are apparently long-lived, and sizeable specimens crown hilltops near Fairfax in Marin County (pl. 56) and in the Santa Lucia Mountains south of Monterey. The madrone is common on the west side of the bay, but much less so on the east side, where it's confined to shaded, north-facing slopes. It ranges into the mountains of southern California and into middle-elevation Sierran pine forests and continues north through the Coast Ranges as far as British Columbia.

MANZANITAS   *Arctostaphylos* spp.

Pls. 57–59

California is home to most of the world's manzanitas, a group of red-barked shrubs primarily adapted to summer-dry climates. Manzanitas range from woody ground covers to small trees that reach up to 25 feet high. The Bay Area is home to many comely species, several of which are rare and restricted to special soils. Manzanitas are identified by their red purple bark (pl. 57); evergreen, usually entire and vertically oriented, ovate leaves; urn-shaped, white or pink flowers (pl. 58); and reddish, applelike berries (pl. 59). Because many manzanitas hybridize, their identification is bewildering to the beginner. Manzanitas are named for their fruits *(manzanita* is Spanish for *little apple),* which were consumed in quantity by the Indians and are widely sought by coyotes, birds, bears, and other wildlife. The genus name is derived from two Greek words meaning *bear grape,* so that both the common and technical names indicate essentially the same thing. The seeds inside manzanita fruits are bone hard and must pass through the digestive juices of an animal gut before they're able to germinate. A few of our local manzanitas reach true tree status: the common manzanita *(A. manzanita)* has pale green leaves and white to pale pink flowers and is widespread in the North Bay and East Bay on the edge of woodlands and forests; the big-

Plate 57. Common manzanita. Manzanitas are noted for their multiple trunks that wear glossy, dark red bark.

berry manzanita *(A. glauca)* has bluish green leaves and white flowers and occurs in the hot chaparral from Mt. Diablo south; the King Mountain manzanita *(A. regis-montana)* has clasping, pale green leaves and white flowers and lives on rocky outcrops in the King Mountain area of San Mateo

Plate 58. Common manzanita. The pale pink, lantern-shaped blossoms of the common manzanita appear in late winter.

Plate 59. Common manzanita. The common manzanita is also noted for its vertically oriented leaves and red berries that ripen in spring, attracting myriad animals.

County. These and a few others grow from 15 to 20 feet tall, usually as multitrunked trees. Some particularly impressive specimens of common manzanita may be seen in Morgan Territory east of Mt. Diablo.

ROSEBAY RHODODENDRON
Pls. 60–62

Rhododendron macrophyllum

The rosebay rhododendron is one of two rhododendrons native to California; the other, known as the western azalea *(R. occidentale),* is a shrub. The rosebay rhododendron varies from a large evergreen shrub to a small tree reaching up to 20 feet tall and occurs along the edges of moist, shaded redwood and other coastal forests (pl. 60). As is so common in the

Plate 60. Rosebay rhododendron. The rosebay rhododendron grows along the edge of moist coastal forests in the extreme North Bay, for example, in the Kruse Rhododendron Reserve on the northern Sonoma coast.

heather family, the rosebay rhododendron grows exclusively on humus-rich, acid soils. Its leaf design, shape, and details of flowers, fruits, and seeds look very much like those of the cultivated rhodies, which originated in the mountains of China and the Himalayas. The leaves are broadly elliptical, tough and leathery, and sometimes curled along the edges at exposed sites (pl. 61). The rose purple flowers are borne in close clusters from late May to early June. Each flower is a broadly flared

Plate 61. Rosebay rhododendron. The glossy, curled leaves of the rosebay rhododendron are large and are the reason for the specific epithet *macrophyllum*.

Plate 62. Rosebay rhododendron. Rosebay rhododendron flowers are large, openly bell shaped, and rose purple with red spots. Look for these flowers in late May.

trumpet, with the petals joined below and sprinkled with dark purple spots (pl. 62). The elongate seed pods start with a velvety covering, then turn dry and brown, and finally split into five starlike sections from which spill hundreds of tiny, dustlike seeds. The rosebay rhododendron is rare in the Bay Area, although there are localized stands as far south as the Santa Cruz Mountains, but it is much more typical of the northern coast, where it lights up forest edges all the way north into Washington State. The best stands in our area occur in Kruse Rhododendron Reserve near Salt Point State Park on the northern Sonoma coast, and in the adjacent Sea Ranch area.

Fabaceae (Pea Family)

One of the world's largest, most varied families, with three distinct subfamilies. All family members have nitrogen-fixing nodules on the roots; usually pinnately compound, pealike leaves with stipules; and pealike seed pods (legumes) characterized by a single row of seeds that opens along two lengthwise seams when ripe. Each subfamily has its own flower design; the Mimosoideae (mimosa, or acacia, subfamily) is noted for tiny petals and bunches of long, colorful stamens; the Caesalpinoideae (senna subfamily) has showy, slightly irregular petals and unfused stamens; and the Papilionoideae (pea subfamily) has a butterfly-like design in which the petals are arranged as a banner (upper back petal), two wings (lateral petals), and a keel (two fused central petals that are boat shaped). In the pea subfamily, usually nine of the 10 stamens are fused into a sheath. The pea family comprises thousands of species and is prominent in every part of the world from the alpine zone of mountains to hot deserts, and from the subarctic to the tropics. It is the second most important family economically, with legume crops (peas, lentils, beans) providing significant amounts of vegetable protein, and forage crops (alfalfa, clover, vetch) vital to livestock. Plants range from tiny annuals to large, rainforest trees. Most of the Bay Area trees in this family are naturalized ornamentals, with only a single native shrub sometimes attaining tree status. The acacias are noted for their fluffy clusters of colorful stamens; the western redbud *(Cercis occidentalis)* for its heart-shaped leaves and rose purple flowers; and the black locust *(Robinia pseudoacacia)* for its stout trunks and white, pea-shaped blossoms.

WATTLE, ACACIA, BLACKWOOD   *Acacia* spp.
Pls. 63–66   NOT NATIVE

Acacias are a widespread group in excess of a thousand species. Most are found in seasonally dry areas of the subtropics and tropics. Some are shrubs, others trees. Most of the aca-

Plate 63. Silver wattle. The silver wattle is an acacia typified by pale, cream-colored blossoms and feathery, twice pinnately compound leaves.

cias we cultivate in our gardens come from Australia, with the result that people often have the impression that all acacias are Australian. Yet California has one native shrubby acacia—catclaw *(A. greggii)*—found in the desert washes of southern California. Several of the Australian acacias have become naturalized in coastal areas, usually on sandy soils. *The Jepson Manual* (Hickman 1993) lists several species to be expected here; I detail the three that you're most likely to encounter. These fall into two categories according to leaf design: those with twice pinnately compound leaves, typified by a distinctly fernlike appearance; and those that have simple, lance-shaped leaves called phyllodes. The only reasonably common pinnately leaved acacia here is the silver wattle *(A. dealbata)*, noted for the dense silvery hairs on its leaves and its ball-shaped heads of bright yellow flowers (pl. 63). Acacias with simple leaves have modified leaf stalks that serve as functional leaf blades; the true leaf blades have been lost, for they would be thinner and more vulnerable to rapid water loss. Phyllodes are thicker and more water conserving, a decided advantage to acacias that inhabit areas with a dramatically long, dry season. Our two acacias with phyllodes are the Sydney golden wattle *(A. longifolia)*, with spikes of bright yellow flowers, and the blackwood *(A. melanoxylon)*, with spherical heads of pale

Plate 64. Blackwood. The sickle-like leaves of the blackwood are held edgewise; the pale yellow flowers are arranged in dense balls.

Plate 65. Blackwood. The black-wood has narrow, lance-shaped phyllodes in place of ordinary leaves.

Plate 66. Blackwood. The crown of the blackwood is broadly rounded, and the branches are thick and dense.

yellow flowers (pl. 64). The Sydney golden wattle seldom tops 20 feet, whereas the blackwood may reach up to 50 feet—a far more substantial tree (pl. 66).

PLUME ALBIZIA or MIMOSA   *Albizia lophantha*

NOT NATIVE

At first glance, the plume albizia looks like another kind of acacia. It is a 20-foot tree with semievergreen, twice pinnately divided leaves; the leaves are deep green and velvety. The semievergreen leaves sometimes remain on the tree through mild winters but are lost when a cold snap hits. Short spikes of yellow green flowers are borne in fluffy masses in the same manner as acacia flowers and appear in late spring. Native to Asia, the plume albizia naturalizes on sandy soils along our coast. A better known relative, *A. jubrissin* ("the" mimosa of eastern U.S. gardens) is better known as an ornamental in our gardens and features pink, powderpuff-like clusters of flowers. This species evidently does not escape to grow wild the way its sister species does.

WESTERN REDBUD   *Cercis occidentalis*

Pls. 67–70

The western redbud is one of the true glories of the Sierra and inner North Coast Ranges foothills, blossoming from late March to April. Although it usually grows as a large, multibranched shrub, treelike forms up to 20 feet high also occur; in gardens they can be pruned to grow in many different shapes. Everything about the western redbud is distinctive: the large, kidney- to heart-shaped, deciduous leaves (pl. 67); the dense clusters of rose purple, sweetpea-shaped flowers that emerge before the leaves in early spring (pl. 68); and the long, deeply rose-colored, beanlike seed pods (pl. 69) that ripen in late spring or early summer. Despite the fact that the flowers give the impression of being true pea blossoms, the arrange-

Plate 67. Western redbud. Western redbud leaves are unusual in the pea family because they're simple and broadly heart to kidney shaped.

Plate 68. Western redbud. The western redbud bears dense clusters of rose purple, sweetpealike flowers in early spring before the leaves open.

Plate 69. Western redbud. The broad, flat, red seed pods of the western redbud are as colorful as the flowers.

Plate 70. Western redbud. Western redbud trees are as broad as they are tall. They often grow in the inner North Coast Ranges foothills with the gray pine *(Pinus sabiniana)*.

ment of the petals is not in the right order, and so the western redbud is placed in the senna subfamily (Caesalpinoideae). The Indians often coppiced the western redbud to encourage straight new shoots for long strips of bark used in weaving fine baskets. It is common only in the extreme northeastern corner of the Bay Area in northern Solano, eastern Napa, and Lake counties in the hot, dry foothill country where gray pine *(Pinus sabiniana)*, interior live oak *(Quercus wislizenii)*, and blue oak *(Q. douglasii)* thrive (pl. 70).

BLACK LOCUST   *Robinia pseudoacacia*

Pls. 71–73   NOT NATIVE

The black locust has long been cultivated in California, often as a street tree in Bay Area cities, as a shade tree next to farm buildings, or occasionally naturalized in the hot, inner foothills, mostly on canyon bottoms. Native to the southeastern United States, the black locust is a stout tree with shallowly fissured, near-black bark; an oval crown; long, deciduous, pinnately compound leaves (pl. 71); racemes of white, sweetpea-shaped flowers (pl. 72); and long, dark brown seed

Above: Plate 71. Black locust. The black locust produces dense branches of bright green, pinnately compound leaves.

Top right: Plate 72. Black locust. The fragrant, hanging trusses of white, sweetpealike flowers of the black locust hang from leafy branches.

Right: Plate 73. Black locust. Because the black locust suckers, a single tree may grow into a tightly knit grove.

pods. Because of its thirsty roots and messy leaves and pods, the black locust is seldom planted in newer landscapes, although the rose-flowered forms are attractive enough that their liabilities may be forgotten. Even if the parent tree is damaged, it readily suckers and will renew its life through them (pl. 73). The black locust also uses a large amount of water, and consequently, little grows underneath it.

Fagaceae (Beech or Oak Family)

Shrubs and trees with simple to lobed, evergreen or deciduous leaves. It has inconspicuous, usually wind-pollinated, unisexual flowers: tiny male flowers are borne on gracefully drooping or stiff, upright catkins; female flowers occur singly or in small clusters in leaf axils or at the base of male catkins. Female flowers are seldom noticed because of their small size; they consist of bracts surrounding one or few ovaries topped by a three-lobed stigma. Ovaries ripen into hard-shelled nuts, which in oaks are called acorns. The nuts are enclosed in spiny bracts or sit in cuplike structures composed of warty or scaly bracts. Three genera occur in the Bay Area: *Lithocarpus* (tanbark-oak), typified by stiff, white male catkins and a bristly acorn cup; *Chrysolepis* (chinquapin), characterized by stiff, white male catkins and nuts surrounded by viciously spiny bracts; and *Quercus* (true oaks), identified by dangling, yellowish male catkins and scaly or warty acorn cups.

COAST CHINQUAPIN *Chrysolepis chrysophylla*
Pls. 74–76

The coast chinquapin (formerly called *Castanopsis chrysophylla*), typically forms colonies or groves (pl. 74), and individual trees are often multitrunked. The coast chinquapin is easily singled out by the golden undersides of its tough, leathery, ovate leaves (pl. 75)—the golden wax protects the leaves from drying out on hot days. Coast chinquapin bark is pale gray to whitish and often decorated with a variety of mosses, lichens, and leafy liverworts. Flowers appear in late spring, with the white candles of the male flowers poking stiffly upright. At the base of these flowers are spiny female flowers with protruding stigmas. The female flowers are conspicuous decorations as they ripen their nutlike fruits in fall; by then, their spiny burrs have enlarged to protect the fruits from browsing animals (pl. 76). Despite this formidable armor, insects often

Plate 74. Coast chinquapin. The coast chinquapin often grows densely in broadly round-topped groves. This one occurs on Montara Mountain near Pacifica.

find a way between the burrs' spines and lay their eggs inside the fruits. Competition for these morsels is keen, because the nuts are edible and sweet. The coast chinquapin comes in two genetically different forms: a tree that reaches over 50 feet high *(C. chrysophylla* var. *chrysophylla)*, and a stiffly branched shrub *(C. chrysophylla* var. *minor)*. The shrub form typifies coastal chaparral on rocky slopes; the trees occur on moist slopes in the fog belt, often in company with the Douglas-fir *(Pseudotsuga menziesii)*, tanbark-oak *(Lithocarpus densiflorus)*, and California hazelnut *(Corylus cornuta* var. *californica)*. Beautiful stands of coast chinquapins can be seen on Montara Mountain near Pacifica and on Mt. Tamalpais.

Above: Plate 75. Coast chinquapin. Here you can see the bicolored nature of coast chinquapin leaves: green above and golden beneath.

Left: Plate 76. Coast chinquapin. Coast chinquapin leaves are glossy and slightly upwardly folded. The spiny burrs make their identification unmistakable.

TANBARK-OAK   *Lithocarpus densiflorus*

Pls. 77–79

The tanbark-oak is not a true oak because of its different male flowers and mode of pollination—beetles visit the stiff, white, upright, ill-scented male catkins (pl. 77). Botanists often classify the tanbark-oak nearer to the chestnuts (*Castanea* spp.) than to the true oaks. The tanbark-oak is a tree that is often found in company with other evergreen trees that seek generous winter rainfall and cool, foggy summers. The trunks may soar to over 100 feet high under ideal conditions, and the gray

Plate 77. Tanbark-oak. The slender, white male catkins of the tanbark-oak sit stiffly erect and are beetle pollinated, whereas those of the true oaks hang and are wind pollinated.

bark becomes deeply fissured with age. The high tannin content of this bark was once utilized for tanning hides and lends the tree its common name. Tanbark-oak leaves reach four to six inches long and are tough, leathery, and lance-shaped, with a coarsely serrated edge and a pinnate vein pattern embossed in the leaf surface. The young leaves are a lovely bronze color (pl.78). The leaves' undersides are covered by a close mat of wooly hairs. Look for the white male catkins in late spring to early summer; at their bases you'll find a few tiny female flowers topped by stubby stigmas. Large, meaty acorns ripen in midfall, tucked into shallow, saucer-shaped cups ornamented

Plate 78. Tanbark-oak. The new leaves of the tanbark-oak are bronze colored. Here they contrast with a common companion tree, the Douglas-fir (*Pseudotsuga menziesii*).

Plate 79. Tanbark-oak. Tanbark-oaks often grow in dense groves with narrow, upright trunks. These trunks appear mottled because of the thick growth of lichens on them.

by many rounded bristles. Tanbark-oak acorns were ground and leached for food by the Indians, along with acorns from true oaks; the tanbark-oak often produces an abundant crop. The tanbark-oak is commonly encountered in the company of the madrone *(Arbutus menziesii)*, Douglas-fir *(Pseudotsuga menziesii)*, California bay *(Umbellularia californica)*, and coast redwood *(Sequoia sempervirens)*. It is common in the Santa Cruz Mountains and in the North Bay but is missing from the generally drier East Bay.

COAST LIVE OAK   *Quercus agrifolia*
Pls. 80–84

The coast live oak is the hallmark of the oak woodlands that cover coastal hills; it also lines canyon bottoms farther inland. It may grow mixed with other oaks or form forests exclusively by itself. Like other tree oaks, the broad, rounded crown may be wider than high (pl. 80). The multiple, muscular trunks (pl. 81) give the trees a decidedly imposing appearance. The dark gray bark is at first smooth but later develops irregular cracks

Plate 80. Coast live oak. Here the dark, rounded crowns of coast live oaks stand out. The smaller, leafless trees are willows.

Plate 81. Coast live oak. The dark gray trunk of the coast live oak is thick and muscular.

and fissures. The coast live oak may be distinguished from other similar-looking oaks by its leaves: leaf margins are curled under to expose less surface to the sun (especially on the upper branches), leaves are dark green on both surfaces (pl. 82), and tiny clumps of whitish hairs occur along the vein

Left: Plate 82. Coast live oak. Coast live oak leaves are broad, glossy, and curled along their edges.

Below: Plate 83. Coast live oak. These fat red buds of the coast live oak contain future male catkins.

junctions on the underside. People often mistake these hairs for some sort of mealy bug, but a hand lens quickly reveals their true identity. In spring, hundreds of red-budded male catkins (pl. 83) open to reveal yellow stamens, emitting clouds of pollen into the wind. Smooth, plump, acorns, often striped, sit in shallow, scaly cups (pl. 84), ripening in October and November. The coast live oak was another favorite food plant of the Indians. Look for it on rolling hills in the Santa Cruz Mountains, much of Marin and Sonoma counties, on flatlands along the eastern margin of the bay, and in canyon bottoms farther inland, as on Mt. Diablo. In many cases it grows in the company of the Douglas-fir *(Pseudotsuga menziesii),*

Plate 84. Coast live oak. These rounded, plump acorns are still green; they'll turn brown by October.

California bay *(Umbellularia californica)*, madrone *(Arbutus menziesii)*, and, in the north, Garry oak *(Q. garryana)* and goldcup oak *(Q. chrysolepis)*. On the Monterey Peninsula, as pitch canker claims the lives of Monterey pines *(Pinus radiata)*, the coast live oak is taking over as the dominant tree.

GOLDCUP OAK or CANYON LIVE OAK   *Quercus chrysolepis*
Pls. 85–87

The goldcup oak has a broad distribution throughout the foothills and lower mountains of California; its native range is from the Klamath Mountains along the Oregon border to the Laguna Mountains in the south of San Diego County. Like the coast live oak *(Q. agrifolia)*, the goldcup oak is often muscularly multitrunked, but the bark is pale gray and on mature trees is displayed as thin strips. Although its crown is often rounded, it may be narrow if it's competing for light with taller trees (pl. 85). The leaves are often confusing to the novice, for some have prickly, hollylike teeth, and others are smooth and entire. But the underside of the leaves quickly identifies this tree: a gold powder covers leaves when they first appear in spring, then later the undersurface turns gray. The upper side of all leaves is dark green, providing a strong contrast (pl. 86). When the fat acorns ripen in fall, they sit in a warty acorn cup covered with more gold powder (pl. 87), giving rise to both the common name and the specific epithet, which translates as *golden scale.* The goldcup oak typifies a small group of oaks, sometimes referred to as golden oaks,

Plate 85. Goldcup oak. Goldcup oak crowns are often as tall as they are broad where they reach for light between taller trees such as these Douglas-firs *(Pseudotsuga menziesii)*.

Plate 86. Goldcup oak. The goldcup oak is readily identified by the decidedly bicolored leaves illustrated in this photo.

Plate 87. Goldcup oak. Goldcup oak acorns are a rich, mahogany brown and sit in warty cups covered with a gold powder.

that is native to the southwestern United States and northern Mexico. Species in this group combine traits from the black oak group with traits from the white oak group. Look for the goldcup oak in canyon bottoms or on rocky promontories; apparently, its persistence there depends on underground water that roots can tap into during drought. This oak is widely distributed in the Bay Area, with some truly massive specimens in the Santa Lucia Mountains of Monterey County; smaller trees crown the summit of Mt. Diablo and the tall hills in Sugarloaf State Park in Sonoma County.

BLUE OAK   *Quercus douglasii*

Pls. 88–90

The blue oak is strongly indicative of the hot inland foothills of the Coast Ranges and Sierra Nevada. Although the crown is often less massive than that of either the coast live oak *(Q. agrifolia)* or the goldcup oak *(Q. chrysolepis)*, the blue oak can reach a height of 60 feet in the northern Sierra and southern Cascade ranges. The blue oak is immediately recognized by its less than three-inch-long, shallowly lobed leaves with a distinctively bluish tint (pl. 88). The blue color comes from a protective waxy covering, which reflects away the most energetic waves of sunlight to keep the leaves cool in summer. In

Left: Plate 88. Blue oak. Blue oak leaves are decidedly blue green and shallowly lobed.

Above: Plate 89. Blue oak. Blue oak acorns sit in warty cups. This acorn is unusual in that it's germinating while still attached to the parent tree.

Plate 90. Blue oak. The blue oak is typical of the inner Coast Ranges foothills. These newly leafed-out trees were photographed at Mitchell Canyon on Mt. Diablo.

addition, blue oak leaves are only shallowly lobed as compared to those of the California black oak *(Q. kelloggii),* valley oak *(Q. lobata),* and Garry oak *(Q. garryana).* The bark is pale gray to whitish and shallowly fissured in a usually vertical pattern of strips. The plumply rounded acorns sit in shallow, warty cups and sometimes sprout while still on the tree (pl. 89). The blue oak is also noted for its wide variety of colorful and fanciful galls—cups, saucers, sea urchins, starbursts, and more. Although other oaks also feature galls, none seem to surpass the diversity of the galls found on the blue oak. Galls are tumorlike growths that are stimulated to form by the tiny stingless wasp larva that lives inside. Eventually, the larva pupates, burrows its way out, and flies away. The blue oak has been unfairly maligned by ranchers because of the belief that its roots take water away from pasture grasses, but in fact, the roots stabilize the soil and help retain more water than if there were no trees on the land. Look for blue oaks on the drier sites of Mt. Diablo (pl. 90), in Sugarloaf State Park and the adjacent Napa Valley, and even on the drier inland sides of the Santa Cruz Mountains on the Peninsula. They often grow in company with the gray pine *(Pinus sabiniana)* and interior live oak *(Q. wislizenii).*

GARRY OAK or OREGON WHITE OAK   *Quercus*
Pls. 91–93   *garryana*

The Garry oak is a common sight in the North Bay, becoming ever more abundant as you go north into Oregon and southern Washington, but it's found only occasionally in the South Bay or East Bay. In details of leaf and bark design—deeply scooped, rounded lobes on the leaves (pl. 91) and checkered,

Above: Plate 91. Garry oak. Garry oak leaves are bright green and deeply lobed.

Left: Plate 92. Garry oak. Garry oak bark is pale gray and shallowly checkered. This particular tree has yellowish crustose lichens growing on its bark.

Plate 93. Garry oak. The Garry oak may form exclusive stands in the North Bay, such as on this slope in Annadel State Park, Sonoma County. Douglas-firs *(Pseudotsuga menziesii)* crown the hilltop.

grayish white bark (pl. 92)—the Garry oak closely resembles its sister species, the valley oak *(Q. lobata)*, but its habit, acorns, and habitat all differ. The Garry oak crown seldom reaches the great dimensions of the valley oak crown, nor do its branches gracefully droop close to the ground. The acorns are rounded and fat and sit in warty cups, whereas those of valley oak are far more slender and sit in warty cups that are short in comparison to the acorn length. The Garry oak is also distinguished in spring by the felted pink hairs on the new leaves, although these hairs are not nearly as conspicuous as those on California black oak *(Q. kelloggii)* leaves. Like the blue oak *(Q. douglasii)*, black oak , and valley oak, the Garry oak is deciduous in late fall—in fact, all California oaks with lobed leaves are deciduous. Look for the Garry oak on parts of Mt. Tamalpais, on Carson Ridge in Marin County, and in Annadel (pl. 93) and Sugarloaf state parks in Sonoma County. Although it may form nearly pure stands where it grows, the Garry oak is sometimes mixed with the California black oak, Douglas-fir *(Pseudotsuga menziesii),* and madrone *(Arbutus menziesii).*

CALIFORNIA BLACK OAK   *Quercus kelloggii*

Pls. 94–97

This particular black oak is only one of four species (the coast live oak *[Q. agrifolia]*, Shreve oak *[Q. parvula* var. *shrevei]*, and interior live oak *[Q. wislizenii]* are the others) that may be found within our borders, and it is the only one that bears the common name *black*. All the other oaks, except for the gold-cup oak *(Q. chrysolepis)*, belong to the white oak group. In general, black oaks have dark, deeply fissured bark (pl. 94); leaves edged with bristle-tipped teeth or lobes (pl. 95); scaly acorn cups; acorn shells that are fuzzy on the inside; and acorns that require two years to ripen. By contrast, white oaks have pale, thinly fissured bark; leaves that lack bristle tips;

Plate 94.
Black oak.
The black oak
is often multi-
trunked and
characterized
by blackish
bark.

Plate 95. Black oak. Black oak leaves are deeply lobed, and each lobe is bristle tipped. Black oak leaves turn shades of gold and orange in fall.

Plate 96. Black oak. The new spring leaves of the black oak are covered with bright pink hairs.

warty acorn cups; acorn shells that are smooth inside; and acorns that ripen in less than a year. The California black oak is deciduous, but in leaf, it's immediately distinguished by the large, deeply lobed leaves whose tips end in conspicuous, slender bristles (pl. 95). In early spring, the brand new leaves wear a vivid pink frosting of hairs (pl. 96) that makes the tree look like an exotic tropical bloomer. The acorns sit in scaly cups and were a favorite food of the Indians because of their rich, oily taste. Like the Garry oak *(Q. garryana),* the California black oak is not particularly common in the East Bay or South Bay areas (you can see it at Morgan Territory and on the top of Mt. Hamilton), but it is a common component of oak woodlands and mixed forests north of the Golden Gate, particularly in

Plate 97. Black oak. These newly leafed-out black oak crowns are tall and narrow because of the intense competition between the closely spaced trees in this woodland.

Sonoma and Napa counties, where it may form nearly exclusive stands. The California black oak is also a common component of middle-elevation mixed-conifer forests throughout California's mountains. Fine old specimens are encountered in Yosemite Valley.

VALLEY OAK or ROBLE   *Quercus lobata*
Pls. 98–100

No other oak in the United States can surpass the size of the immense, broadly rounded crown of the valley oak. Although oaks in general seldom live more than 200 years, some well-known valley oaks have been aged at close to 300 years. True

Plate 98. Valley oak. The valley oak has a broadly rounded crown with drooping branches that sometimes almost touch the ground.

to its name, the valley oak seeks valley bottoms and well-watered sites, although it seldom lives where soils remain soggy through summer. Instead, look for the valley oak on deep alluvial plains or on flats with a high water table. The valley oak once covered thousands of acres of the Sacramento and northern San Joaquin valleys, but agriculture has made short shrift of most, and only marginal forests, often near watercourses or as windbreaks along roads, remain. The valley oak is best distinguished by its silhouette: a broad rounded crown with the end branches gracefully drooping and sometimes touching the ground (pl. 98). The valley oak is also identified by its checkered whitish bark (pl. 99); the large, deeply lobed leaves (pl. 100); and the relatively long, tapered acorns. The acorns and overall shape, not to mention habitat, distinguish it from its close relative the Garry oak *(Q, garryana)*. Despite the size of the trunks on many valley oaks, their wood was seldom used, for it is disappointingly weak and rots easily, earning it the name *mush oak*. Perhaps the best stand of valley oaks is in a broad savanna on the military

Plate 99. Valley oak. The valley oak is characterized by a massive trunk covered by finely checkered or fissured, pale gray bark.

Plate 100. Valley oak. Valley oak leaves are dark green and deeply lobed.

property of the Hunter Liggett Reservation just southwest of King City. There are many remnant stands of valley oaks scattered throughout the Bay Area, and some excellent trees occur on the plains just east of Sebastopol in Sonoma County.

INTERIOR LIVE OAK   *Quercus wislizenii*

Pls. 101–103

The interior live oak is the counterpart to the coast live oak *(Q. agrifolia);* it usually lives inland on the hot, dry foothills surrounding the Central Valley, often in company with the blue oak *(Q. douglasii)* and gray pine *(Pinus sabiniana).* Consequently, the interior live oak is far more drought and heat tolerant than its close relative, and as a result, it is often of much smaller stature and seldom becomes an impressive tree. The interior live oak can be distinguished from the coast live oak by its leaves, which are dark green on both sides, flat (the

Plate 101. Interior live oak. In spring, the interior live oak may produce prodigious numbers of reddish-tinted male catkins.

margins don't curl under as they do on the coast live oak), and lack hairs on the underside. Interior live oak leaf margins vary from completely smooth to prickly toothed, whereas those of the coast live oak all bear sharp teeth. Where the ranges of the two oaks overlap, as near Napa Valley and on Mt. Diablo, the two species sort themselves out by habitat: the coast live oak seeks canyon bottoms, whereas the interior live oak grows on the sides and tops of hills (pl. 103). An interesting stand of

Plate 102. Interior live oak. All oaks produce tiny, axillary female flowers. Here you see one at the base of the middle leaf in the photograph.

Plate 103. Interior live oak. From a distance the dark green, broadly rounded crown of the interior live oak looks similar to that of the coast live oak.

goldcup oaks *(Q. chrysolepis)* and interior live oaks crowns the summit of Mt. Diablo, where you can walk through a short forest along the Fire Interpretive Trail. One other live black oak merits mention: the Shreve oak *(Q. parvula* var. *shrevei).* Superficially, the Shreve oak resembles the interior live oak except that its leaves may be twice as long (up to four inches), the undersides are drab olive green instead of bright green, and the acorns are abruptly narrowed at their tips rather than gradually tapered from the middle to the tip. The Shreve oak is generally poorly known and is scattered in moist mixed forests in the Santa Cruz Mountains.

Garryaceae (Garrya Family)

Large evergreen shrubs or small trees with tough, leathery, ovate, opposite leaves. The plants are dioecious; Male flowers are borne in long, narrow, dangling catkins—each male flower having boat-shaped bracts and long stamens. Female flowers are carried in similar but shorter catkins, also having boat-shaped bracts and pistils bearing long, dark stigmas. The ovary ripens into a fleshy, berrylike fruit with red purple pulp. This small family consists of the single genus *Garrya* (silk-tassel bushes). Only the coast silk-tassel *(G. elliptica)* reaches tree status in the Bay Area.

COAST SILK-TASSEL   *Garrya elliptica*
Pls. 104, 105

The coast silk-tassel can grow as a densely branched shrub or as a multi- to single-trunked tree up to 25 feet high. Truly large specimens have trunks more than a foot in diameter. A first impression of the coast silk-tassel bush is that it's some kind of manzanita: the leaves and stature are surprisingly similar. But the coast silk-tassel's leaves are opposite with wavy margins (pl. 104), and the bark is gray or brownish with

Plate 104. Coast silk-tassel. Coast silk-tassel leaves are leathery, dark green, and oval and have a fluted edge.

Plate 105. Coast silk-tassel. In winter, the slender, hanging male catkins of the coast silk-tassel put on a dramatic display even though they're wind pollinated.

no hint of red. When the coast silk-tassel blooms near the end of winter, there is no mistaking its identity: long, gray green male catkins (pl. 105) and shorter female catkins adorn separate shrubs. Despite the showiness of these catkins, the flowers are petal-less and wind pollinated. Grapelike, grayish-husked fruits ripen in summer and are poisonous to humans. The coast silk-tassel is often cultivated for its handsome foliage and striking catkins. It occurs naturally on Montara Mountain near Pacifica, at Huckleberry Preserve in the Berkeley Hills, along the Sonoma coast, and in coastal forests of Marin County and the Peninsula.

Hippocastanaceae (Horse-chestnut Family)

Deciduous shrubs or trees with opposite, palmately compound leaves. White, yellowish, red, or pink flowers are borne in dense, candlelike clusters. Flowers have five sepals, five somewhat irregular petals, five stamens, and a pistil with a superior ovary. Fruits are large one- to three-chambered capsules with one to three large, round, chestnut-shaped seeds. This small Northern Hemisphere family consists mainly of one genus: *Aesculus* (buckeye, horse-chestnut), with only one species native to California and the Bay Area.

CALIFORNIA BUCKEYE **Aesculus californica**

Pls. 106–110

The California buckeye is a familiar sight on hills and along canyon bottoms, especially in the inner Coast Ranges (pl. 106) and Sierra foothills, but it is also encountered in more coastal forests, sometimes mixed with the California bay *(Umbellu-*

Plate 106. California buckeye. California buckeyes and oaks often grow together. In this photo, the buckeyes are on the lower parts of the hills and have just leafed out; coast live oaks *(Quercus agrifolia)* crown the hilltops.

Plate 107. California buckeye. The leafless, whitish branches of the California buckeye are in stark contrast to those of other trees in the winter landscape.

Plate 108. California buckeye. California buckeye leaves are palmately compound and form a delicate, green tapestry in mixed forests in early spring.

laria californica), the madrone *(Arbutus menziesii),* and live oaks. The California buckeye is a low, widely spreading tree with a broad, sometimes mushroom- or cauliflower-shaped crown. It is also a tree that changes with the seasons: bare, sil-

Left: Plate 109. California buckeye. California buckeye flowers are borne in candlelike clusters in late spring. Notice the protruding stamens.

Above: Plate 110. California buckeye. California buckeye fruits look like leathery pears before opening.

ver gray bark glows in winter light (pl. 107); new, apple green leaves appear in early spring (pl. 108); candles of white to pale pink, fragrant flowers (pl. 109) light up forests and woodlands in late spring; and leathery, pear-shaped seed pods (pl. 110) decorate trees in late summer and fall. The glossy brown seeds are reminiscent of chestnuts and lend the common name *horse-chestnut,* but beware of these, for they are highly poisonous. The Indians made use of these poisons, crushing the seeds and tossing them in streams to stupefy fish. They also cooked and leached the seeds for a starchy, edible mush in times of poor acorn harvests. The tree had other uses, too, especially the twigs, which were made into pointed fire drills. The California buckeye is found almost throughout the Bay Area, but especially large specimens occur on the water district land of the Crystal Springs area in San Mateo County, and a curiously shaped, near-sprawling, wind-pruned tree stands near the end of the Bear Valley Trail in Point Reyes National Seashore.

Juglandaceae (Walnut Family)

Deciduous trees with alternate, resinously scented, pinnately compound leaves. The wind-pollinated, unisexual flowers are borne on the same tree: fat, rounded male flowers in hanging catkins; squat female flowers in clusters of two or three at the base of the male catkins. Female flowers have an inferior ovary topped with two broadly crested stigmas. Fruits are fleshy-hulled, nutlike drupes containing a single, large, usually edible seed that is often called a nut. Nuts of commercial value include the butternut, walnut, hickory nut (*Carya* spp. and *Juglans* spp.), and pecan *(C. illinoiensis)*. California has a single species with two varieties: *Juglans californica* var. *californica* (southern black walnut) is found in the coastal canyons of southern California; *J. californica* var. *hindsii* (northern black walnut) is our local version. Sometimes the English walnut *(J. regia)* also persists in abandoned fields and along watercourses, and so it is included here.

NORTHERN BLACK WALNUT   *Juglans californica*
Pls. 111, 112   **var. *hindsii***

The northern black walnut is noted for its rich, oily nutmeats and thick shells, which are hard to crack without pulverizing the nutmeats inside. For this reason, the northern black walnut is seldom cultivated for food, but it serves as the vigorous rootstock onto which the English walnut *(J. regia)* is grafted, and when a walnut orchard has been abandoned, the rootstocks may sprout as suckers and overshadow the grafted top. The northern black walnut is considered a tree of limited natural distribution; one of the original locales was Walnut Creek. The Indians were fond of these nuts and also used their shells for games and gambling, so they often planted them in canyons throughout the central and north-central Coast Ranges, as well as in the Sierra foothills. So when you're on an outing in a canyon, you may chance upon a small grove of

Plate 111. Black walnut. Black walnut leaves are pinnately compound with several pointed leaflets.

northern black walnuts that has persisted for many years. The black walnut is seldom noted for its blossoms, but the petalless, wind-pollinated flowers appear just as the new leaves are issuing from their buds in midspring. The shiny, resinous leaves, with 11 or more leaflets per leaf (pl. 111), are conspicuous from early spring to fall when they turn soft shades of buttery yellow (pl. 112). You can easily observe black walnuts

Plate 112. Black walnut. In fall, the arching trunks and crowns of pale golden leaves of the black walnut stand out.

at the mouth of Mitchell Canyon on Mt. Diablo and in the main canyon in Sugarloaf State Park.

ENGLISH WALNUT or PERSIAN WALNUT *Juglans regia*
Pls. 113–115 NOT NATIVE

The English walnut is the preferred nut crop of this genus and is widely planted in California's foothills, especially where there is a high water table. Although Walnut Creek was originally home to the native northern black walnut *(J. californica* var. *hindsii)*, the English walnut was also once widely grown there before suburbanization paved over the orchards. You can easily distinguish the two walnuts. The northern black walnut has near-black bark, whereas the English walnut has paler gray bark. The northern black walnut has several pointed leaflets per leaf, whereas the English walnut has only five to nine broad, rather rounded leaflets per leaf (pl. 113).

Plate 113. English walnut. English walnut leaves have fewer, broader, more-rounded leaflets per leaf than do black walnut leaves. These leaves display fall color.

Plate 114.
English walnut.
English walnuts
bear female flow-
ers crowned with
two furry-looking,
broad stigmas to
trap wind-borne
pollen.

Plate 115.
English walnut.
By contrast, the
male flowers of
English walnuts
are arranged in
thick, drooping
catkins. These
haven't opened
yet.

English walnuts are preferred for food because their shell is thinner and easier to crack, and the nutmeat inside is larger, although it is not as richly flavored. Like our native black walnut, the English walnut has female flowers crowned with two furry-looking stigmas (pl. 114) and has dangling male catkins (pl. 115). Only occasionally will you encounter English walnuts growing "wild," but there are places where orchards have been abandoned and the trees persist on their own.

Lauraceae (Laurel Family)

A large, mainly tropical family of shrubs and trees with simple, leathery, alternate, often evergreen leaves full of aromatic oils. Small greenish, yellow, or whitish flowers are borne in small clusters, each usually with three layers of perianth parts in which the distinction between sepals and petals is not clear. The flower parts are generally in multiples of three, and the stamens often have hoods or appendages over their pollen-producing anthers. The superior ovary ripens into a fleshy, one-seeded drupe, which, in several species, is edible by humans. This family is noted for the avocado *(Persea americana),* the bay *(Laurus nobilis),* and the tree that produces cinnamon *(Cinnamomum zeylandicum),* among others. California has a single, widespread tree species: the California bay *(Umbellularia californica).* We also plant the camphor tree *(C. camphoratum)* along streets in the Bay Area.

CALIFORNIA BAY or PEPPERWOOD or OREGON MYRTLEWOOD   *Umbellularia californica*

Pls. 116–118

The California bay is richly represented in several of California's forests, including the mixed-evergreen forests of the Coast Ranges, the mixed-conifer pine belt of the mountains, the riparian woodlands of southern California, and the edges of redwood forests. The California bay seems to thrive on both floodplains and rocky hill tops, as long as there's sufficient summer water to carry the trees through to the fall rains. The California bay has a highly malleable shape (pl. 116); it may soar to 100 feet when competing for light, grow as a shrub on serpentinite soils, or sprawl as a wind-pruned tree of great character. It may be single-trunked but is often multi-trunked when the crown has been damaged. The deep green, lance-shaped leaves are highly scented, sometimes overpoweringly so, but judicious use of the leaves adds flavor to stews, soups, and sauces. Small yellow flowers (pl. 117) appear any-

Plate 116.
California bay.
These California bay
crowns have
assumed varied shapes
according to
the direction
and severity
of the local
winds.

Plate 117.
California bay.
Here you see
the leathery,
narrowly
ovate leaves
and clusters
of pale yellow
flowers of the
California bay.
The flowers
usually open
during winter.

Plate 118. California bay. California bay fruits look like miniature avocados and are still green in early fall. Eventually, they'll turn dark purple.

where between late November and March and provide one of the few nectar sources for pollinators during that slow time of the year. Small, avocado-like fruits swell by summer's end but remain green (pl. 118), finally turning deep purple just before falling. The fruits are sought out by birds and mammals. The Indians made use of the roasted ground seeds as a condiment; animals often eat the fleshy covering. You can find the California bay in all but the driest parts of the Bay Area.

Moraceae (Mulberry Family)

Mostly tropical plants, the majority of which are shrubs or trees with simple, sometimes lobed, alternate leaves that contain a milky sap. Minute, green, petal-less unisexual flowers are borne in dense clusters, are sometimes wind pollinated, and are found either on the same or different plants. Ovaries ripen into drupes or achenes, which are sometimes embedded in a fleshy receptacle to form a syconium (figlike fruit). None are native to California, but mulberries (*Morus* spp.) and the edible fig *(Ficus carica)* are sometimes naturalized. Only the latter is to be expected growing naturally in the Bay Area. The family also has other prominent edible plants such as the jackfruit and breadfruit (*Artocarpus* spp.).

EDIBLE FIG
Pls. 119, 120

Ficus carica
NOT NATIVE

The Spanish missionaries introduced the edible fig into California with the first missions. Popular as a fruit in Spain and originating in the Middle East, it adapted well to California's similar climate. The edible fig often persists around old ranches and farms and also occasionally grows naturally in the canyons of inland foothills (pl. 119). The edible fig is immediately recognized by its palmately lobed leaves spanning up to six inches across (pl. 120), with milky sap. The flowers

Plate 119. Edible fig. Mature edible fig trees have a broad crown. Here the leaf canopy is backlit.

Plate 120. Edible fig. Edible fig leaves are nearly round in outline and deeply palmately lobed.

are never in evidence because they're borne on the inside of a fleshy receptacle that later ripens into the purple, edible fruit called the fig. The fig receptacle (the syconium) is hollow on the inside and lined with hundreds of tiny, greenish male and female flowers. Even more curious, the edible fig depends on tiny stingless wasps that force their way in through a tiny hole at the end of the syconium. The female wasp lays her eggs inside, and when they hatch, the male wasps fertilize the females. As the female wasps move about they get covered with pollen, exit the syconium, and fly to a new fig and start the process all over. Most modern edible fig varieties are self-fertile and so don't require the services of the wasps, but early varieties needed them and failed to set fruit until the right wasps had been imported.

Myoporaceae (Myoporum Family)

Perennials, shrubs, or small trees with simple, evergreen leaves dotted liberally with dark glands (hold a leaf up to the light). Flowers range from slightly to strongly two-lipped, each with five petals (two upper and three lower) joined to form a tube, and four stamens and a superior ovary that usually ripens into a fleshy berry. This small family is closely related to the large snapdragon family (Scrophulariaceae) but differs in its glandular leaves and fleshy fruits. It is mainly confined to Australia and the Pacific islands. Few members of the family are cultivated in California except for the genus *Myoporum,* of which one species is sometimes naturalized as a small tree along the coast.

NEW ZEALAND MYOPORUM *Myoporum laetum*
Pls. 121–123 NOT NATIVE
The New Zealand myoporum is often planted as a fast-growing shrub or small tree; it may develop a single trunk when appropriately pruned (pl. 121). It is well adapted to

wind, salt in the air, and sandy soils and makes an excellent choice for ocean-front gardens. Birds sometimes eat the purplish berries and spread the seeds to wild sites, where the New Zealand myoporum appears to grow on its own. The New Zealand myoporum is distinguished by its oval, translucent, dark green leaves liberally sprinkled with black glands (pl. 122); and it produces axillary clusters of small, white flowers

Plate 121. New Zealand myoporum. New Zealand myoporum trunks may grow tall, with narrowly fissured bark.

Plate 122. New Zealand myoporum. New Zealand myoporum leaves look speckled when the light passes through, because of dark glands embedded in the leaves.

Plate 123. New Zealand myoporum. The New Zealand myoporum has narrowly oval leaves and clusters of small, white, slightly irregular flowers.

with purple spots sprinkled on the lower lip (pl. 123). Other myoporums are sometimes cultivated—particularly the little-leaf myoporum *(M. parvifolium)*, a prostrate ground cover—but they never appear beyond the confines of gardens and parks.

Myricaceae (Sweet-gale Family)

Shrubs or small trees with simple, alternate, aromatic, deciduous or evergreen leaves. Small, greenish, unisexual flowers are borne in dense axillary clusters or in short catkins, sometimes on separate plants. The ovaries ripen into waxy, warty drupes. This small family has a broad distribution, and two species are native to California: the California wax-myrtle *(Myrica californica)*, or bayberry, occurs in the Bay Area.

CALIFORNIA WAX-MYRTLE or BAYBERRY　　　***Myrica***
Pls. 124–126　　　　　　　　　　　　　　　　　　***californica***

The common name *bayberry* is unfortunate, for it suggests a relationship with the California bay *(Umbellularia californica)*, which is not the case. In fact, the eastern bayberry *(M. cerifera)* is of minor fame for the fragrant candles that are made from the wax boiled from its berries. Our species, also called the California wax-myrtle, has too low a wax content to be useful. The California wax-myrtle is a lover of forest edges, growing quickly into a hedge, dense shrub (pl. 124), or

Plate 124. California wax-myrtle. New branches of the California wax-myrtle often grow stiffly in vertical files, as seen here.

Right: Plate 125. California wax-myrtle. Female trees are laden with bumpy, dark purple fruits in summer and fall.

Below: Plate 126. California wax-myrtle. The California wax-myrtle produces small clusters of wind-pollinated flowers in leaf axils. The dark green leaves are narrowly lance shaped and toothed.

slender tree up to 30 feet tall next to closed-cone pine and red-wood forests, in habitats that receive plenty of summer fog. Its rapid growth and handsome foliage have made the California wax-myrtle increasingly popular in Bay Area gardens. The California wax-myrtle can be identified year round by its glossy, spicily aromatic, evergreen leaves that are edged with tiny teeth and often wavy along the edges (pl. 125). Its odor is much more subtle than the potent smell of the California bay and is most noticeable on warm spring or summer days. The insignificant flowers are seldom observed (pl. 126), but the dark purple fruits become conspicuous by summer's end (pl. 125). Look for vigorous stands of California wax-myrtle at Salt Point State Park, along the coastal strip of Big Basin State Park, or among coastal forests in Pt. Reyes National Seashore.

Myrtaceae (Myrtle Family)

Evergreen shrubs to large trees with alternate or opposite, simple, entire, strongly aromatic leaves. The leaf odor is often reminiscent of camphor. Flower designs vary: some species have five conspicuous petals; others have tiny or no petals. Most species have many to numerous, often colorful stamens attached along the edge of a nectar-secreting disk. The inferior ovary ripens into a woody, several-chambered capsule or becomes a fleshy berry. This large family is about equally divided between the American tropics and the Asian tropics and Australia. Many of the species with woody capsules have diversified in the Land Down Under. Many different shrubs and trees of the family are cultivated in California, including the myrtle *(Myrtus communis)*, tea-trees *(Leptospermum* spp.), bottlebrushes *(Callistemon* spp.), paperbarks *(Melaleuca* spp.), and of course, the far-flung eucalyptuses *(Eucalyptus* spp.; 600 species in Australia). Allspice and cloves also belong to this family. Only eucalyptuses are to be expected as naturalized trees in the Bay Area, and I detail here two of the most commonly encountered species.

RIVER RED GUM *Eucalyptus camaldulensis*

Pls. 127, 128 NOT NATIVE

Eucalyptuses come in many forms, from dense, low shrubs to the world's tallest flowering trees, but they all have in common a singular flower and fruit structure: sepals and petals are fused into a cap that falls from the flower bud as it opens; the numerous stamens attract pollinators (stamens may be white, yellow, red, or pink); and the woody seed pods are covered with a cap that falls away when seeds are ripe. In fact, the name *eucalyptus* means *well or truly covered* in Greek. Most mature

Plate 127. River red gum. The muscular trunks of this eucalyptus have mottled bark.

Plate 128. River red gum. The river red gum bears broadly rounded crowns. These particular trees were photographed in their Australian homeland.

eucalyptuses are further characterized by their vertically oriented lance- to sickle-shaped leaves. Some of the important distinguishing features among eucalyptus species are the bark patterns, arrangement of flowers, details of the flower bud's cap, and details of the fruit. The majority of eucalyptuses feature flowers with white stamens, as do the two described here (a few are notable for their red, pink, or yellow stamens). The river red gum is noted for its smooth, grayish brown bark that peels in strips (pl. 127); its five- to 15-flowered umbels; and a bud cap shaped like an inverted half circle with a nipplelike beak in the center. It is widely naturalized in disturbed, mostly coastal areas from the northern coast through the Bay Area and on southward. It reaches a large size on bottomlands in its native habitat (pl. 128).

BLUE GUM

Eucalyptus globulus

Pls. 129–132

NOT NATIVE

Most Californians refer to this tree as "the" eucalyptus, hardly realizing that this species is but one of hundreds. The blue gum was the tree introduced in the late 1800s as a fast-growing wonder tree that would provide a cheap and convenient source of lumber. Many eucalyptuses, in fact, are sources of excellent quality wood, but this tree's wood is too dense and hard to be easily worked. It is also a thirsty tree, for despite the fact that it grows in many hot, dry places in California, it comes from the moist, sclerophyll forests found in southeastern Australia. The blue gum survives here by its extensive, deeply delving roots. The blue gum is so named because the broad leaves of juvenile trees (pl. 129) are liberally covered with a blue powder, whereas the adult leaves are green, sickle shaped (pl. 129), and vertically oriented. Sometimes the contrast between young and old trees is particularly striking because of these

Plate 129. Blue gum. The blue gum has entirely different juvenile and adult leaves. Broad, bluish leaves are found on young trees or stump sprouts, whereas hanging, sickle-shaped leaves characterize the adult foliage.

Above: Plate 130. Blue gum. There is a sharp contrast between the young saplings and the mature trees of the blue gum.

Left: Plate 131. Blue gum. The blue gum is characterized by pendulous branches and peeling strips of bark.

Plate 132. Blue gum. Often, the best way to identify the blue gum is to look for dropped, single, white blossoms and blue, bell-shaped seed pods.

differences (pl. 130). The blue gum features smooth, pale bark that peels in long strips (pl. 131) (in this respect it is not terribly different from the river red gum) and a bud cap that is bluish and heavily wrinkled (the basic shape is like that of the red river gum). Like most eucalyptuses, its flowers are white, but the seed pods turn bluish until they ripen (pl. 132). The blue gum is still widely grown, especially as a windbreak and living fence, and it has spread from many of its early plantings. It is especially common in the Oakland and Berkeley hills, where it may cover entire hillsides with a barren-looking forest. Even though cold snaps appear to kill the trees, the roots are tough and vigorously stump sprout when mild conditions return.

Oleaceae (Olive Family)

Vines, shrubs, and trees, often with opposite leaves that may be simple or pinnately compound. Flower design varies widely from greenish, petal-less, and wind-pollinated to showy and insect-pollinated. Those with petals usually have four joined into a tube, and there are usually only two stamens. The superior ovary ripens into a fleshy drupe, an achene, or a winged samara. This modest-sized family has a broad distribution, especially in the drier parts of the Northern Hemisphere, although the ashes (*Fraxinus* spp.) are important riparian trees. Several ornamentals belong here, including jasmines (*Jasminum* spp.), lilacs (*Syringa* spp.), and forsythias (*Forsythia* spp.); and the olive *(Olea europea),* from the Middle East, is an important source of food and oil. California has a few native species in the desert, and the ashes, which occur typically along canyon bottoms and streams. In the Bay Area, we occasionally see naturalized olives, and we encounter two species of ash, noted for their singly-winged samara-type fruits.

FLOWERING ASH   *Fraxinus dipetala*

Pls. 133–135

The flowering ash is a large shrub or small tree attaining 15 to 20 feet in height. Like other ashes, it is a water lover, seeking canyons and protected, north-facing slopes in woodlands, where its roots are able to tap into moisture, even in summer. Like other ashes, it features opposite, pinnately compound leaves, although most of its leaves have only three to five leaflets (pl. 133), a rather low number and probably a reflec-

Plate 133. Flowering ash. Flowering ash leaves are characterized by having only three to five leaflets per leaf; other ashes have more.

Plate 134. Flowering ash. The white flower trusses of the flowering ash are dense and showy, attracting bees for pollination.

Plate 135. Flowering ash. These unripe, green flowering ash samaras bear only a single wing, as opposed to the double wings of maple samaras.

tion of the fact that this ash lives in places where summers are hot. Unlike most other North American ashes, the flowering ash is insect pollinated and has two white petals per flower. The flowers are massed into branched spikes that open in April (pl. 134), and because of the quantity of flowers produced, the result is a showy display. Typical, pendant, one-winged samaras (pl. 135) ripen in summer and are carried on winds to new sites. The flowering ash is scattered through the Bay Area; look for it along Putah and other creeks in the inner North Coast Ranges, in the Donner Canyon area of Mt. Diablo, and in Pinnacles National Monument. Its chief distribution lies in the Sierra foothills and South Coast Ranges.

OREGON ASH   *Fraxinus latifolia*
Pls. 136–138

The Oregon ash is much more typical of the genus in North America than is the flowering ash *(F. dipetala)*: it is a tree up to 70 feet tall that is partial to permanent watercourses; the pinnately compound leaves have seven or more leaflets per leaf (pl. 136); and the tiny, greenish, wind-pollinated flowers are borne with male and female flowers on separate trees. The flowers appear just as the leaves are bursting from their buds in early spring (pl. 137). The male flowers look like dense bunches of long stamens; the female flowers consist of pistils with sticky stigmas. The one-winged samaras that ripen in summer are far more likely to attract attention, but perhaps

Plate 136. Oregon ash. The Oregon ash is characterized by pinnately compound leaves with several pointed leaflets per leaf.

the most striking feature is the fall color provided by the yellow leaves (pl. 138). The Oregon ash is scattered in the South Bay and West Bay, and reaches its best development in the north; fine stands can be seen in Sugarloaf and Annadel state parks in Sonoma County, and along the Navarro River in Mendocino County.

Above: Plate 137. Oregon ash. The dense clusters of male, wind-pollinated flowers of the Oregon ash open just as the new leaves are emerging.

Right: Plate 138. Oregon ash. The rounded crowns of the Oregon ash turn to fallow gold in fall.

OLIVE
Olea europea

Pls. 139–141 `NOT NATIVE`

The olive tree is highly symbolic and has been revered for millennia in the Middle East and Mediterranean regions of the Old World, where its fruit and oil have provided a bountiful source of food. Well adapted to hot, dry summers, the olive was one of the first trees introduced into California by the Spaniards. Since then, it has been widely planted for its beauty as well as its fruits and has become a major cash crop in California's Central Valley. As with so many other well-adapted nonnatives, the olive occasionally escapes cultivation when birds carry off its seeds. The olive is noted for its fluted and gnarled bark; broadly spreading crown (pl. 139); narrowly

Plate 139. Olive. The broad, rounded crown of an olive tree is reminiscent of the crowns of California's oaks.

Plate 140. Olive. Olive leaves are whitish beneath, dark green above.

Plate 141. Olive. Olives are characterized by narrowly lance-shaped leaves and clusters of tiny, white flowers.

lance-shaped, grayish leaves (pl. 140); clusters of small, whitish, petal-less flowers (pl. 141); and black, fleshy, intensely bitter fruits. In order to make olives edible, they have to be treated with lye! Some olive cultivars don't fruit heavily, for many gardeners find the fruits more of a nuisance than a benefit.

Pinaceae (Pine Family)

Mostly evergreen trees with needlelike leaves. Pollen and seed cones are borne on the same plant. Seed cones have papery or woody, spirally arranged scales that carry two seeds each on their upper surface. Cones range from an inch to two feet long. When the needles are shed, they fall separately from their twig or are attached to minute spur shoots, whereas in the redwood family the twig falls with the needles attached. This prominent conifer family is confined to the Northern Hemisphere and is richly represented in California. Our native genera, most of which occur in the Bay Area, include *Pinus* (pines), with one to five needles borne on spur shoots; *Abies* (firs), with needles that leave a smooth scar behind when they fall, and seed cones that are carried upright and shatter when ripe; *Picea* (spruces), with needles that leave a distinct peg behind, and hanging seed cones that fall intact; *Tsuga* (hemlocks), with needles that leave a roughened place behind, and hanging seed cones that fall intact; and *Pseudotsuga* (Douglas-fir), with needles similar to those of the hemlock, and seed cones that fall intact and have prominent three-pronged bracts between the scales.

SANTA LUCIA FIR — *Abies bracteata*
Pls. 142–144

True firs are immediately distinguished from other members of the pine family—including the well-known Douglas-fir *(Pseudotsuga menziesii)*—by their upright, candlelike seed cones that appear on the uppermost branches of the tree, well out of reach. The seed cones fall to pieces when they've ripened and turned brown. You're unlikely to see a whole, intact cone on the ground, but rather the individual pieces: dozens of scales and seeds. The Santa Lucia fir is noted for its long, pitch-laden bracts that extend between the scales of the cones, glistening in the sun and giving this tree its alternate

Plate 142. Santa Lucia fir. The bristly looking seed cones of the Santa Lucia fir are produced near the top of the crown. These cones shatter when they fall.

name: bristlecone fir (pl. 142). This fir is also distinctive for its narrow, steeple shape and its stout, spine-tipped needles (pl. 143). The Santa Lucia fir is California's rarest fir, occurring only on hot, dry limestone scree in the northern Santa Lucia Mountains (pl. 144), where it lives perched high above the wild Pacific Ocean south of Monterey. The easiest way to see this fir is to take Nacimiento Road near the coastal town of Lucia to the

Plate 143. Santa Lucia fir. The Santa Lucia fir is unusual among firs for having stiff, spine-tipped needles.

Plate 144. Santa Lucia fir. The shape of a Santa Lucia fir is narrow and steeplelike. Here you see a tree in its native habitat near Cone Peak in the Santa Lucia Mountains.

mountains' crest, then turn north onto Cone Peak Road. Cone Peak, whose summit is reached by a windy trail, is home to Santa Lucia firs, sugar pines *(Pinus lambertiana),* and Coulter pines *(P. coulteri).*

GRAND FIR   *Abies grandis*

Pls. 145–147

The grand fir clings closely to the northern coast from the Salt Point area of northern Sonoma County northward, usually within sound of the surf (pl. 145). It often joins company with the Douglas-fir *(Pseudotsuga menziesii),* Sitka spruce *(Picea*

Plate 145. Grand fir. Where grand fir trees are exposed to extreme winds, the branches lie flat and the trees assume a compact shape. These grand firs are growing at Salt Point State Park along the northern Sonoma coast.

Plate 146. Grand fir. The lower branches of the grand fir are typified by glossy, blunt needles arranged flat in two rows.

Plate 147. Grand fir. Grand fir seed cones are cylinder shaped and sit upright on the branches like fat candles.

sitchensis), and western hemlock *(Tsuga heterophylla),* often along the western edge of the redwood belt. The grand fir is indeed grand, often topping 100 feet in height, with stiff, formal-looking whorls of branches along its trunk; two flat rows of glossy, deep green, blunt-tipped needles (pl. 146); and slender, cylinder-shaped seed cones (pl. 147). Unlike the Santa Lucia fir *(A. bracteata),* its cones have no obvious bracts between the scales. In addition to these features, a close perusal reveals a tiny, V-shaped notch at the end of each needle. The grand fir barely enters the northwestern fringe of the Bay Area but becomes increasingly abundant northward into Mendocino, Humboldt, and Del Norte counties. Some dramatically wind-pruned grand firs may be seen near Stump Beach in Salt Point State Park.

KNOBCONE PINE   *Pinus attenuata*

Pls. 148–151

The knobcone pine belongs to the closed-cone pine group: species that retain their seed cones for life—try removing one and you'll see how tightly attached the cones are. This feature,

and the fact that the cones remain tightly closed until the tree dies (pl. 148), adapts these pines to periodic fire (pl. 149), allowing the forest to be newly occupied by thousands of newly sprouted seedlings in the aftermath. For this reason, most knobcone pines in a stand are of the same age and look remarkably uniform. The knobcone pine is distinguished by its rela-

Top: Plate 148. Knobcone pine. The tapered seed cones of the knobcone pine open only after fire or when a branch has been severed.

Left: Plate 149. Knobcone pine. Even when they've been burned, knobcone pines retain their seed cones.

Above: Plate 150. Knobcone pine. The knobcone pine bears slender needles in threes. These branches are accompanied by many developing pollen cones.

Left: Plate 151. Knobcone pine. Knobcone pine bark is irregularly fissured.

tively short needles in threes (pl. 150), classical conical shape when young, irregularly fissured bark (pl. 151), and scrawny appearance when mature. Its asymmetrical, tightly held seed cones appear on all the upper branches and extend down the trunk. Each seed cone has knoblike protrusions on the outer side (pl. 148). The knobcone pine is adapted to nutrient-poor, rocky soils where summer temperatures soar; look for it in the Mt. St. Helena area and near Knobcone Point on Mt. Diablo. An interesting stand occurs on the lithified sand dunes near Bonny Doon in the Santa Cruz Mountains, near the rare Santa Cruz cypress.

BEACH PINE   *Pinus contorta* subsp. *contorta*

Pls. 152–155

Strictly speaking, the beach pine doesn't come within the boundaries of the nine Bay Area counties, but it does occur just to the north along the rugged Mendocino coast. This is a pine with great variability: the coastal variety is restricted to the high winter rainfall and fog belt along the immediate coast northward into Washington, whereas its sister variety,

Plate 152. Beach pine. Here you see a whorl of purple, baby seed cones around the time that they'll be pollinated. It takes nearly two years for the seeds inside to ripen.

the lodgepole pine *(P. contorta* subsp. *murrayana),* is one of the most abundant high-elevation conifers in western North America. The tree shapes of these two varieties are quite different: the beach pine has a broad, rounded crown (pl. 154) and seldom tops 40 feet, whereas the lodgepole pine develops a narrow crown atop trunks that may reach close to 100 feet. Lodgepole pine bark is covered by small, cornflake-shaped scales, which are lacking in the beach pine. But when it comes to needles and seed cones, the two varieties are almost identical: both feature two twisted needles per spur shoot, and small

Plate 153. Beach pine. Clusters of pollen cones alternate with the short, bushy needles of the beach pine.

Plate 154. Beach pine. The beach pine wears a broad, rounded crown.

seed cones that seldom exceed three inches in length. The scales of the seed cones are narrow and tipped with a slender, inturned prickle. One other difference between the two varieties is that the beach pine may retain its seed cones for several years, but the lodgepole pine always sheds its mature seed cones. A third form of this pine—*P. contorta* subsp. *bolanderi,*

Plate 155. Beach pine, Bolander form. The Bolander pine occurs in pygmy forests along with the pygmy cypress (*Cupressus goveniana* subsp. *pygmaea*). Although a subspecies of beach pine, it differs in that it retains its seed cones for a long time.

or Bolander pine—is noted for its dwarf habit, the slow opening of its seed cones (pl. 155), and its survival on the ancient marine terraces where the pygmy cypress *(Cupressus goveniana* subsp. *pygmaea)* grows in dwarfed form.

COULTER PINE   *Pinus coulteri*
Pls. 156–158

The Coulter pine counts as one of three native pines with especially large, heavy seed cones; the cones may weigh more than three pounds each. These seed cones are dangerous when they fall not only because of their weight, but because their scales end in a wickedly stout, upturned spine (pl. 156). The other two pines with similar cones are the gray pine *(P. sabini-*

Plate 156. Coulter pine. The massive, heavy seed cones of the Coulter pine weigh more than those of any other pine. Each scale ends in a wicked, upturned spine.

Plate 157. Coulter pine. The densely tufted, bluish green needles of the Coulter pine help distinguish it from the gray pine (Pinus sabiniana), which has sparse, gray green needles.

ana), described below, and the Torrey pine *(P. torreyana),* from far coastal southern California. Part of the weight of these cones is due to the large, meaty, highly edible seeds. The seeds are important food for jays and squirrels and were once gathered by the Indians who lived in the Coast Ranges. The Coulter pine might be confused with the gray pine, but its needles in three are bushier and denser, not quite so gray, and don't droop (pl. 157). The Coulter pine's seed cones are somewhat longer than broad, rather than equally long and broad. The Coulter pine reaches its northern limits in the Bay Area at Black Diamond Mines Regional Park near Antioch; other stands are localized in Mitchell Canyon on Mt. Diablo, and near the top of Mt. Hamilton (pl. 158). The Coulter pine is most abundant from the Santa Lucia Mountains south into

Plate 158. Coulter pine. Coulter pine trees crown the rocky upper slopes of Mt. Hamilton.

the mountains of northern Baja California, where it often grows with the ponderosa pine *(P. ponderosa)*, Jeffrey pine *(P. jeffreyi)*, or big-cone Douglas-fir *(Pseudotsuga macrocarpa)*.

SUGAR PINE   *Pinus lambertiana*
Pls. 159–162

Although the Coulter pine *(P. coulteri)* may have the heaviest seed cones, the sugar pine has the longest. Each cylinder-shaped seed cone spans from one to two feet in length (pl. 159). But sugar pine cones are light, with papery, nonprickly

Plate 159. Sugar pine. The seed cones of the sugar pine are immediately recognizable by their length, shape, and lack of prickles.

Plate 160. Sugar pine. The horizontally trending branches of the sugar pine carry their long, slender seed cones at the branch tips.

Plate 161. Sugar pine. The massive trunk of an old sugar pine carries its branches at right angles.

scales, and they produce sweetly flavored, winged seeds carried away by wind or by hungry squirrels. Because the cones are so distinctive, you can easily identify the sugar pine from a distance, for the seed cones are carried at the ends of branches and can be seen hanging like long pendants (pl. 160). In addition, the sugar pine's silhouette stands out: the main branches diverge stiffly at right angles from the trunk (pl. 161). The sugar pine is named for its sugary sap that sometimes oozes

Plate 162. Sugar pine. Sugar pine needles are slender and relatively short.

from the bark, but it belongs to a group of pines known as white pines. White pines are valued for their wood, and many species occur around the Northern Hemisphere—two others occur in California's high mountains. White pines, in addition to having prickleless, cylindrical, hanging seed cones, are characterized by slender needles borne in fives (pl. 162). The sugar pine is uncommon in the Bay Area—it reaches its best development in the middle elevations of the Sierra in company with the ponderosa pine *(P. ponderosa)*, incense-cedar *(Calocedrus decurrens)*, and white fir *(Abies concolor)*—but it occurs in small stands in the higher mountains. Look for it on Mt. St. Helena and in the Santa Lucia Mountains south of Monterey.

BISHOP PINE *Pinus muricata*
Pls. 163–166

The bishop pine is the second in a series of three closed-cone pines found in the Bay Area (see also the knobcone and Monterey pines [*P. attenuata* and *P. radiata*]). Closely related to the Monterey pine, it differs by having slightly bluer green needles borne in twos (not threes) (pl. 165), and smaller seed cones up to about four inches long. But the habitat and overall size and shape of the crown—which varies according to wind conditions—are similar to those of the Monterey pine.

Plate 163. Bishop pine. Bishop pine seed cones are firmly attached for life along the branches. With age, they sometimes open.

Plate 164. Bishop pine. The pollen cones of the bishop pine are carried in long, dense clusters below the new needles.

Both pines were also considered relicts, but recent fossil evidence indicates that their populations have fluctuated widely according to periods of greater drought or heavier rainfall. Both pines now live along sea bluffs in areas that receive heavy summer fog; the fog condenses on needles and falls to the ground as summer rain. Both pines are adapted to rocky or sandy, nutrient-poor soils where other trees can't compete (pl. 166). But the curious thing is that the bishop pine and the

Right: Plate 165. Bishop pine. Bishop pine needles are dense and often more bluish than those of the Monterey pine (*Pinus radiata*).

Below: Plate 166. Bishop pine. Here you see the dramatic difference the soil makes in determining vegetation: the bishop pines in the distance are growing on decomposed granite, the coastal scrub in the foreground is growing on soils of sedimentary origin.

Monterey pine seldom meet: the bishop pine's best development occurs north of the Golden Gate, on Inverness Ridge, and along the northern Sonoma and Mendocino coasts. The two pines do overlap in one fascinating botanical area known as Huckleberry Reserve near Seventeen-Mile Drive on the Monterey Peninsula.

PONDEROSA PINE   *Pinus ponderosa*

Pls. 167–170

Like the Douglas-fir *(Pseudotsuga menziesii)*, the ponderosa pine is one of the most widespread conifers in the western United States. It typifies the broad, middle-elevation forest belt in most of California's mountains, especially in the Sierra Nevada, where it grows with the sugar pine *(P. lambertiana)*, white fir *(Abies concolor)*, incense-cedar *(Calocedrus decurrens)*, and sometimes Douglas-fir. The ponderosa pine starts out with a fairly narrow shape, but as it matures the crown broadens and the major limbs often swoop down in a ponderous way (pl. 167). Perhaps no other California pine is more "typical" of pines in general: the bushy needles are clear green and in threes (pl. 168), and the seed cones are perfectly symmetrical, each scale ending in a slender, outturned prickle (pl. 169). If you remember that *p* stands not only for *ponderosa* but *prickly,* the association is easy, for if you're in the mountains,

a second closely related conifer— the Jeffrey pine *(P. jeffreyi)* — looks similar but has cone scales with inturned prickles ("gentle Jeffrey").

Plate 167. Ponderosa pine. The mature ponderosa pine has "ponderous" looking branches that swoop down.

Plate 168. Ponderosa pine. The thick, luxuriant, bright green needles of the ponderosa pine are especially striking on young trees.

The mature ponderosa pine is noted for its handsome, jigsaw-puzzle-like bark made of small, honey-colored plates (pl. 170). The ponderosa pine has an odd distribution in the Bay Area: as might be expected, most stands occur on the higher mountains such as Mt. St. Helena and Mt. Hamilton, but unusual

Plate 169. Ponderosa pine. The symmetrical seed cones of the ponderosa pine have out-turned prickles on their scales, whereas Jeffrey pine *(Pinus jeffreyi)* cones have inturned prickles.

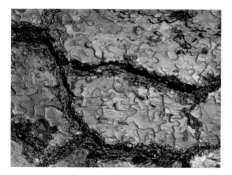

Plate 170. Ponderosa pine. The bark on a mature ponderosa pine looks like a fanciful jigsaw puzzle.

populations also occur in Henry Coe State Park east of the Santa Clara Valley (where the ponderosa pine and gray pine [*P. sabiniana*] sometimes grow together), and in the sandhills of the Santa Cruz Mountains near Felton and in Henry Cowell State Park.

MONTEREY PINE   *Pinus radiata*
Pls. 171–173

The Monterey pine is the most famous of the closed-cone pines and perhaps the best known of all of California's pines, for though it is rare in nature, it is widely planted as a fast-growing timber tree in Chile, New Zealand, Australia, and other distant places. Whereas Californians received the Australian blue gum, the Aussies acquired our "radiata" pine. In California, the Monterey pine is found in only three coastal locales: the Cambria-Hearst Castle area on the southern Monterey coast, the Monterey Peninsula, and the southern San Mateo–northern Santa Cruz coast near Año Nuevo State Park. In all three places, it grows on soils derived from sandstone, shale, and volcanic rocks, within sight of the surf. Although it is adapted to conditions similar to those of the even rarer Monterey cypress *(Cupressus macrocarpa)*, it is usually found slightly farther from the cliff edge than the cypress is. Together with the Monterey cypress, the Monterey pine has made the

Plate 171. Monterey pine. These old seed cones of the Monterey pine are finally opening as they dry out. You can tell their extreme age by the lichens growing on the scales.

Plate 172. Monterey pine. The bark of a mature Monterey pine has deep, vertical fissures.

Plate 173. Monterey pine. Here you see an overmature stand of Monterey pines at Pt. Lobos State Reserve. Notice the old trees are showing symptoms of poor health, but in the open spaces new saplings are growing well.

windbreaks of Golden Gate Park possible, and Monterey pines have been widely—some would say too widely—planted throughout coastal California. For example, no Monterey pines are native to the Berkeley and Oakland hills, and yet today they form dense forests there. Although these nonnative stands of Monterey pine are pleasant on hot days, they have excluded many of the native shrubs and wildflowers, and in the process lowered the area's biodiversity. Unfortunately, although the Monterey pine is handsome and fast growing, it is highly flammable, and so not particularly suitable for most residential areas. Like other closed cone pines, the Monterey pine's seed cones open only after fire or when they've been severed from the parent tree; cones may show their great age by providing a home for colorful lichens (pl. 171). Also, like the bishop pine *(P. muricata),* the Monterey pine's bark is vertically fissured (pl. 172). Short-lived, these pines often show signs of poor health in old age (pl. 173).

GRAY PINE or FOOTHILL PINE   *Pinus sabiniana*

Pls. 174–177

The gray pine is an exclusively California pine whose main distribution is on the hottest, dry foothills encircling the great Central Valley. There it often joins company with the blue oak *(Quercus douglasii)*, interior live oak *(Q. wislizenii)*, and California buckeye *(Aesculus californica)*. This big-cone pine is easily identified by its wispy, drooping, grayish needles in

Plate 174. Gray pine. Here you see the wispy, grayish, drooping needles of the gray pine decorated with clusters of pollen cones.

Plate 175. Gray pine. The gray pine often bears more than one main trunk. The tree you see here is growing on a hot, dry, exposed slope near the Geyers in Sonoma County.

Above: Plate 176. Gray pine. The massive and heavy seed cones of the gray pine are about as broad as they are long, whereas Coulter pine seed cones tend to be somewhat longer than they are broad.

Left: Plate 177. Gray pine. The drooping, wispy needles of the gray pine glow when they're backlit by the sun.

threes (pl. 174); its tendency to produce two or more trunks (pl. 175); and its huge, nearly globe-shaped seed cones (pl. 176). Each cone weighs three or more pounds, has scales that end in a stout spine, and has large edible seeds that are important to wildlife for food. The seeds have a flavor comparable to the best pine "nuts" from the desert's pinyon pines. Look for the gray pine around the Mt. Diablo area, along the eastern side of the Napa Valley, in Henry Coe State Park, and along Mines Road south of Livermore. Although many deride the gray pine as not looking very pinelike, it is a beautiful tree when the needles are backlit by the sun (pl. 177).

DOUGLAS-FIR   *Pseudotsuga menziesii*
Pls. 178–181

Not a fir at all, the Douglas-fir has its own genus. In fact, the genus name translates as *false hemlock,* and indeed, the Douglas-fir is most closely allied to the true hemlocks. The bark on large specimens is irregularly fissured into V-shaped patterns (pl. 178), the needles are arrayed all around the branches (pl. 179), and the seed cones hang from the branch tips. Each seed cone has broadly rounded scales; long, narrow, three-pronged bracts ("mouse tails") (pl. 181); and two winged seeds per scale. The cones fall intact and are often sought by rodents for food. The new growth on a Douglas-fir is particularly distinctive, for the new needles are a vivid yellow green and redolent of lemon when crushed. The Douglas-fir is a variable and widespread tree of several habitats, and it is found not only through many of California's mountains but

Plate 178. Douglas-fir. Douglas-fir trunks often become massive with age and develop irregularly fissured bark. Here, the woody stem of an old poison-oak is winding around the base of the tree.

Plate 179. Douglas-fir. Douglas-fir branches are clothed all the way around with needles that are a vibrant yellow green when new.

Plate 180. Douglas-fir. Douglas-fir pollen cones dangle below the branch among the needles.

also north into British Columbia and east through the Rocky Mountains into Mexico. In the Bay Area, it may form nearly pure stands near the coast—the Bear Valley Trail at Pt. Reyes National Seashore and the Crystal Springs watershed in San Mateo counties are good examples—or, more often, it mixes with other evergreen trees, including the madrone *(Arbutus menziesii)*, California bay *(Umbellularia californica)*, tanbark-oak *(Lithocarpus densiflorus)*, and California-nutmeg *(Tor-*

Plate 181. Douglas-fir. The seed cone of the Douglas-fir is unmistakable: the scales are broadly rounded and are interrupted by narrow, three-pronged bracts.

reya californica), and is also a component in slope redwood forests. It is one of the most valuable commercial timber trees in the United States, currently of premier importance for solid wood and laminated beams for construction.

WESTERN HEMLOCK   *Tsuga heterophylla*
Pls. 182–184

The western hemlock, like the grand fir *(Abies grandis),* barely enters the northwestern corner of the Bay Area, occurring occasionally in the moist, coastal forests near Sea Ranch. From there it continues north into British Columbia and east to Glacier National Park in Montana. Although hemlocks closely resemble spruces, their growing tips are gracefully arched, and the main side branches repeat that pleasant pattern (pl. 182). Also, hemlock bark is vertically fissured. The western hemlock is identified by its short, pale green needles of two lengths arranged in two rows (pl. 183), and these seldom exceed two inches in length. Finally, the papery seed cones are the tiniest in the pine family (pl. 184), seldom measuring more than an inch long, and these are arrayed along the branches like so many brown bells. Companion trees include the Douglas-fir *(Pseudotsuga menziesii),* coast redwood *(Sequoia sempervirens),* and bishop pine *(Pinus muricata).*

Plate 182. Western hemlock. The side branches of the western hemlock droop gracefully downward.

Plate 183. Western hemlock. Western hemlock needles are short and of two lengths on each twig.

Plate 184. Western hemlock. The tiny seed cones of the western hemlock have thin, papery scales.

Pittosporaceae (Pittosporum Family)

Evergreen vines, shrubs, or small trees with simple, leathery, usually entire leaves. Showy flowers are borne in cymes or umbels with five separate sepals, five petals that form a loosely knit tube, and five stamens. Fruits are capsules with sticky, colorful seeds or fleshy berries. This small family is widely distributed throughout the Pacific islands and southern Asia and is especially well represented in Australia. Several ornamentals appear in California gardens, including the blue bell creeper *(Sollya heterophylla)*, Australian frangipani *(Hymenosporum flavum)*, and various pittosporums. None are native, but some species of *Pittosporum* escape or are naturalized in coastal areas.

VICTORIAN BOX, TOBIRA, OTHERS

Pittosporum **spp.**

Pls. 185–187

`NOT NATIVE`

Pittosporums are widely planted in the Bay Area for their handsome and durable evergreen leaves and their cream-colored, fragrant flowers. This fragrance travels far in the evening and is reminiscent of the odor of orange blossoms. Most pittosporums grow as dense shrubs and make excellent hedges when they're closely pruned, but some regularly attain tree status. For example, Victorian box *(P. undulatum)* is a small tree that escapes from cultivation near the coast. This eastern Australia tree is a staple in many California gardens and is identified by its wavy, glossy, narrowly elliptical leaves (pl. 185) and its extremely fragrant flowers (pl. 186). A second species, *P. tobira* (tobira), from Japan, may occasionally reach the size of a small tree and is recognized by its broadly spoon-shaped, deep green leaves whose margins curl under, and its pale yellow flowers (pl. 187).

Left: Plate 185. Victorian box. The tiered branches of the Victorian box are laden with small, fragrant cream-colored flowers.

Middle: Plate 186. Victorian box. Victorian box leaves are glossy and wavy. The bell-shaped, cream-colored flowers are sweetly fragrant.

Bottom: Plate 187. Tobira. The tobira bears headlike clusters of highly fragrant flowers and glossy leaves with curled edges.

Platanaceae (Plane Tree Family)

Deciduous trees with alternate, palmately lobed leaves, reminiscent of maple leaves and bearing collarlike stipules. Tiny, greenish, petal-less, unisexual, wind-pollinated flowers are borne in globe-shaped, ball-like clusters on dangling stems; male and female flowers are on the same tree. Female flowers ripen into spiky, ball-shaped, compound fruits that split apart into single-seeded achenes that are spirited away on wind. This tiny, Northern Hemisphere family is associated with riparian areas and has a broad distribution. California has a single native species, which is found at several sites in the East Bay and South Bay.

WESTERN SYCAMORE   *Platanus racemosa*
Pls. 188–190

The western sycamore is a prominent, multitrunked tree favoring riparian woodlands, usually growing where summers are hot. Although its leaves have the palmately lobed pattern typical of maples, it is not related and is easily distinguished

Plate 188. Western sycamore. The large, maplelike leaves of the western sycamore are alternate (maples are opposite), and they bear conspicuous stipules at their base.

by the fuzzy texture of its alternately arranged leaves, the collarlike stipules (which later slide down the twig) (pl. 188), and the hollow petiole base inside which the associated axillary bud hides. (Other trees have axillary buds that are clearly visible above the leaf node and are not hidden from view.) In addition, the bark on the mature western sycamore consists of an irregular jigsaw puzzle of white, tan, and gray pieces (pl. 189); the color is determined by the age and layer of the bark. The ball-shaped flower clusters appear in April just as the leaves are being renewed; the male flowers have protruding yellow stamens, whereas the female flowers bear fuzzy, dark red stigmas. Later, these clusters of female flowers turn into spiny balls (pl. 190). The western sycamore is most often seen in the South Bay, as at Henry Cowell State Park on the San Lorenzo River, and in the East Bay around parts of Mt. Diablo and Morgan Territory, and along Mines Road south of Liver-

Plate 189. Western sycamore. The massive trunk of the western sycamore carries a crazy puzzle pattern of white, tan, and gray bark.

Plate 190. Western sycamore. These spiny balls are nearly ripe clusters of the fruits of the western sycamore. When they turn brown, the balls separate into dozens of one-seeded sections.

more. It continues south, where it is prominent in many riverine situations throughout southern California, and it is also important in the Sierra foothills.

Rhamnaceae (Buckthorn Family)

Deciduous or evergreen shrubs or small trees with simple, alternate, or opposite leaves that bear stipules (stipules often fall soon after leaves develop). Many small flowers are massed together in complex clusters of various shapes. Many species have colorful petals as well as colored sepals, but some species lack petals altogether. Besides the usually five sepals and petals, they have five stamens, a nectar-bearing disk, and a three-sided superior ovary. Fruits are fleshy capsules or drupes. The buckthorn family has a broad distribution, often occurring in arid climates. California has a large number of species, particularly in the genus *Ceanothus* (wild lilac, deerbrush), whose diversity is centered in the state. All ceanothuses have fragrant flowers; colorful blue, purple, pink, or white sepals and petals; and a capsulelike fruit. By contrast, the buckthorns (genus *Rhamnus*) have odorless, greenish yellow flowers and fleshy drupes. A couple of our local ceanothuses reach tree status and are included below. Occasionally, too, the coffee berry *(Rhamnus californica)* may grow up to 15 feet tall, but it has not been included in this book.

JIMBRUSH · *Ceanothus oliganthus* var. *sorediatus*

Pls. 191, 192

The jimbrush forms dense stands of interlacing branches in the chaparral of the inner Coast Ranges, particularly from Mt. Diablo south, where it may reach tree size near canyon bottoms. A part of the chaparral and oak woodland communities, it grows rapidly to maturity after fire, aided by the

Plate 191. Jimbrush. Jimbrush leaves are glossy and have three main veins extending from the base toward the tip.

nitrogen-fixing nodules on its roots. The small, oval, bright green leaves are alternate on greenish twigs (pl. 191), adorned in early spring by rounded clusters of pale bluish flowers (pl. 192). The flowers are intensely fragrant (some claim the odor is like sweet corn tortillas) and a sure draw to bees. The flowers are followed by smooth, three-sided capsules that split

Plate 192. Jimbrush. The jimbrush is covered with hundreds of rounded clusters of pale blue flowers in early spring.

open by spring's end. Common companions to the jimbrush include the common manzanita *(Arctostaphhylos manzanita)*, mountain-mahogany *(Cercocarpus betuloides)*, and toyon *(Heteromeles arbutifolia)*, all of which may also reach small-tree status. In addition, a second species of shrubby ceanothus called the buckbrush *(Ceanothus cuneatus)* is a common companion, but it is quickly distinguished by its whitish flowers and opposite, dark green leaves. The buckbrush never becomes a tree.

BLUE BLOSSOM *Ceanothus thyrsiflorus*
Pls. 193–195

Although the jimbrush *(C. oliganthus* var. *sorediatus)* is common inland, the blue blossom is a typical shrub or small tree in the fog belt, especially near the coast. In the most favored sites, such as along Inverness Ridge in Pt. Reyes National Seashore, the blue blossom becomes a 20-foot-tall treelet, whereas in the harshest, windiest places, such as the bluffs near the Pt. Reyes lighthouse, it is a sprawling, woody ground

Plate 193. Blue blossom. Blue blossom leaves are similar to those of the jimbrush.

cover. The blue blossom is characterized by angled greenish twigs; elliptical, dark green leaves (with three prominent veins running the length of each leaf) (pl. 193); and dense panicles of sky blue flowers (pl. 194) produced in early spring. As with other ceanothuses, the flowers have the aroma of sweet corn tortillas and are immensely popular with bees. The

Plate 194. Blue blossom. The blue blossom is smothered in dense panicles of sky blue flowers in early spring.

blossoms also lather in water as a fine soap. It produces dark purple fruits (pl. 195) that later dry and split open. The blue blossom can also be seen on Montara Mountain near Pacifica and on San Bruno Mountain south of San Francisco. In the North Bay, it has sprung up in great quantities following the 1990s fire on Inverness Ridge in Point Reyes National Seashore. In the East Bay, look for the blue blossom near Jewel Lake in Tilden Regional Park.

Plate 195. Blue blossom. Blue blossom fruits turn a dark, burnished purple color before drying and splitting open.

Rosaceae (Rose Family)

Diverse family of herbaceous plants, shrubs, and small trees with a variety of different leaf forms and shapes. Flowers range from tiny to large and showy; most look like single roses. Flowers usually possess five sepals and five petals, and many to numerous stamens attached to a distinct hypanthium cup. There is great variation in the pistils: some species have several separate pistils with superior ovaries; others have a single pistil with an inferior ovary; still others have some intermediate condition. Ovaries ripen into single-seeded, dry achenes, fleshy drupes and pomes, or complex aggregations of drupelets and accessory fruits. This large family is best adapted to temperate and subarctic climates throughout the world, especially in the Northern Hemisphere. California has many species, including several different shrubs that dominate many habitats, and a few small trees. The trees in this book are mostly large shrubs that occasionally attain tree status: the mountain-mahogany *(Cercocarpus betuloides)* is noted for its petal-less flowers and plumed achenes; wild plums and cherries *(Prunus* spp.) are identified by the tiny glands at the leaf bases and their drupe-type stone fruits; the western hawthorn *(Crataegus suksdorfii)* is known for its thorny side branches and pome-type fruits; the toyon *(Heteromeles arbutifolia)* has tough, serrated, evergreen leaves and fiery red pomes; service berries *(Amelanchier* spp.) are characterized by deciduous leaves that are toothed on the upper half, and by red purple pomes; and cotoneasters *(Cotoneaster* spp.) are recognized by their small, tough leaves and bright red pomes. Cotoneasters —native to eastern Asia—are included here because they are occasionally naturalized in coastal forests.

SERVICE BERRIES $\qquad$ *Amelanchier* spp.

Pls. 196–197

Service berries are so named because a southeastern species ripened its fruits around the time of a major church service. The small, red purple, applelike pomes are edible, if somewhat mealy, and are said to make excellent jelly. These deciduous shrubs or small trees may be a sprawling ground cover on windy sites or soar up to 20 feet tall in protected canyons. Deciduous in winter, service berries are identified the rest of the year by broadly elliptical, bluish green leaves edged with coarse teeth on their upper half (pl. 196). Narrow racemes of snowy white, apple-blossom-like flowers (pl. 197) appear in midspring, briefly enlivening the scene wherever service berries grow. Seldom common in the Bay Area, service berries

Plate 196. Service berry. Service berry leaves are recognizable by their bluish green color and by teeth restricted to the upper half of each leaf.

are widely scattered in openings of mixed-evergreen forests and oak woodlands. Look for them on San Bruno Mountain, in Annadel State Park, or in Las Trampas Regional Park near Danville. They are much more common farther north and quite abundant in the upper parts of the coniferous forests of the Sierra Nevada. Distinction between the two closely related species, the Utah service berry and the alder-leafed service berry *(A. utahensis* and *A. alnifolia),* is blurred and difficult for the amateur to make.

Plate 197. Service berry. The stiff branches of the service berry carry dense clusters of snowy white, apple-blossom-like blooms in mid-spring.

MOUNTAIN-MAHOGANY   *Cercocarpus betuloides*
Pls. 198–200

The mountain-mahogany varies from a multibranched shrub to a tall, slender tree with multiple trunks, sometimes reaching up to 25 feet tall. Like so many members of the rose family, the mountain-mahogany is adapted to summer-dry conditions in the hot chaparral, and it may enter openings in oak woodlands and mixed-evergreen forests, as well. Not related to the mahogany of commerce (which is a tropical tree belonging to the Meliaceae), the mountain-mahogany is named for its hard, durable wood. Identifying traits include the smooth, pale grayish bark, and the small, semideciduous, wedge-shaped leaves that are often toothed on the upper half (like those of the service berries) and bear an embossed,

Plate 198. Mountain-mahogany. Mountain-mahogany leaves are toothed and show a distinctly embossed pinnate vein pattern.

featherlike vein pattern (pl. 198). The leaves closely resemble those of the sister genus *Holodiscus* (creambushes), but the latter has soft-textured leaves that release a fruit scent when crushed. In blossom, the mountain-mahogany is unexceptional, with umbel-like clusters of greenish, petal-less, saucer-shaped flowers borne in the leaf axils in spring (pl. 199). Even though mountain-mahogany flowers are not showy, bees visit them for the nectar held in their shallow hypanthiums. The

Plate 199. Mountain-mahogany. Flowers are small and hidden under the leaves. Although these flowers lack petals, they're a strong draw to bees because of their abundant nectar.

Plate 200. Mountain-mahogany. The mountain-mahogany is more conspicuous in its fruiting stage than during flowering. Here you see the white, plumelike styles protruding from the young fruits.

mountain-mahogany is far more conspicuous in its fruiting phase, when, in summer, the single-seeded achenes carry long, white, plumed styles that sail away on wind (pl. 200). A shrub in full fruit glows in the afternoon light. Look for the mountain-mahogany in the more inland parts of the Bay Area; it has a broad distribution throughout and continues northward and southward.

COTONEASTERS

Pls. 201, 202

Cotoneaster **spp.**

NOT NATIVE

Cotoneasters were once highly popular large shrubs or small trees planted in Bay Area gardens for their vividly colored orange to red, pomelike "berries." (pl. 201) Birds are fond of gorging on these fruits; these bird visits are responsible for the spread of seeds to natural sites in coastal woods. There are several species of cotoneaster, all originally from the mountains of China; some are low and sprawling, whereas others grow up to 15 feet tall. Most species are characterized by small, ovate, often thick, entire leaves; masses of small, white, single, roselike flowers (pl. 202); and brightly colored fruits in fall. Incidentally, the name is not pronounced "cotton-easter," but rather "coh-tone-ee-aster"!

Plate 201. Cotoneaster. The veiny leaves of the cotoneaster contrast with the vivid red "berries"—in botanical language, actually pomes, or little apples.

Plate 202. Cotoneaster. The cotoneaster shows its relationship to roses in its small, single-rose-like, white blossoms.

WESTERN HAWTHORN   *Crataegus suksdorfii*
Pls. 203, 204

European hawthorns are popularly cultivated small trees in parts of the Bay Area, but few gardeners are acquainted with our own native species. The western hawthorn reaches its southern limits in the northern Bay Area; look for it, for example, in the Sebastopol area of Sonoma County and on other moist, meadowy sites close to the coast, where it is typically associated with fertile valleys and riparian woods. It may grow as a multitrunked shrub or small tree up to 20 feet high (pl. 203). As its common name suggests, the western hawthorn has side branches that end in spines, which is the botanical definition of a thorn. Like other hawthorns, its thin, deciduous leaves are irregularly lobed and coarsely toothed

Plate 203. Western hawthorn. The western hawthorn's shape bulges out in the middle and gradually tapers to the top.

Plate 204. Western hawthorn. Western hawthorn leaves are characterized by their deep green color and irregular teeth. Here you also see the young, developing fruits.

(pl. 204), and the short clusters of white blossoms make a brief appearance in late spring. Small, dark purple, pome-type fruits follow at summer's end, attracting birds, but humans find the fruits tasteless and mealy. Sadly, the aggressive European species *C. monogyna* has recently gained a strong roothold in the watersheds of Crystal Springs Reservoir. It differs from our native in that its leaves have several deep, slashlike lobes, and it bears bright red fruits.

TOYON or CALIFORNIA-HOLLY *Heteromeles arbutifolia*
Pls. 205, 206

Toyon is the name conferred by the Spaniards on this large shrub or small tree, whereas the name *California-holly* reflects the resemblance of its late-ripening, bright red orange fruits (pl. 205) to those of the English holly *(Ilex aquifolium)*. The toyon is a stout-trunked plant with hard wood; narrow, elliptical, four- to five-inch-long leathery, coarsely serrated leaves; and pyramidal cymes of small, white, single, roselike blossoms (pl. 206) borne in early summer. The real show comes in late fall, when the fruits ripen. If a bumper crop of these "berries" is produced, many will be left behind by the migrating birds that feed on them, and they will persist into the following spring. An example of a colorful spring composition pairs toyon fruits and clematis flowers. The toyon seldom reaches its greatest size in the Bay Area, although it has no lack

Plate 205. Toyon. Here you see the tough, serrated leaves and masses of red pomes on the toyon in late fall.

Plate 206. Toyon. The tiny, white flowers of the toyon are borne in elongated clusters in summer.

of habitats: it occurs on the edge of mixed-evergreen and oak woodlands, in canyon bottoms in the chaparral, and even in coastal situations. But to appreciate its full potential, a visit to coastal southern California is necessary; there, toyons have trunks and crowns comparable to those of live oaks. In fact, the town of Hollywood is named for the large toyons that once grew there.

NONNATIVE PLUM

Pls. 207, 208

Prunus cerasifera

NOT NATIVE

Whether the nonnative plum you see got there because a bird carried the seed or because someone tossed a pit is immaterial; this plum finds its way into many natural sites throughout the Bay Area. You'll generally encounter it in open woods, often along streamsides, where the extra water tides it over through the driest summers. If you've ever seen a typical garden plum tree, the nonnative plum looks very much like it: it's a small tree with many short, somewhat crooked branches; serrated, deciduous, elliptical leaves; and showy clusters of snowy white blossoms in early spring (pl. 207). Each flower resembles a single rose—the five sepals, five separate petals,

Plate 207. Nonnative plum. In spring, the nonnative plum is covered in snowy white blossoms.

Plate 208. Nonnative plum. Nonnative plum flowers are characterized by five widely spreading white petals and numerous protruding stamens.

and bunches of stamens all sit on top of a narrow hypanthium cup inside of which the pistil resides (pl. 208). As with other plums, the egg-shaped fruits ripen to a dull red or red purple hue. Don't expect the best quality from the fruit of the nonnative plum, although its flesh is altogether edible and makes excellent jelly.

HOLLY-LEAF CHERRY   *Prunus ilicifolia*
Pls. 209, 210

The holly-leaf cherry represents another part of the large genus *Prunus,* one in which the leaves are tough, shiny, and evergreen. Similar species, including the laurel-cherry *(P. laurocerasus)* and the Portuguese-laurel *(P. lusitanica),* from the

Plate 209. Holly-leaf cherry. The hollylike leaves and tight racemes of white, plum-blossom-like flowers help identify the holly-leaf cherry.

Plate 210. Holly-leaf cherry. The large bunches of deep red purple fruits of the holly-leaf cherry are tempting to many animals.

Mediterranean region, are often grown in California gardens. Our own holly-leaf cherry features broadly elliptical leaves with coarsely saw-toothed edges, narrow racemes of cream-colored flowers that appear in mid- to late spring (pl. 209), and large, dark red purple cherries that ripen in late summer (pl. 210). Unlike most other native prunuses, the holly-leaf cherry has fruits with a pleasantly sweet pulp, but the pulp is so thin that you'd have to process many fruits in order to obtain a meal. Instead, the Indians took advantage of the large pits by roasting and grinding them into a flour. Like most prunus pits, holly-leaf cherry pits contain a toxin that turns into hydrogen cyanide when chewed; the same chemical is also present in the leaves and is identified by the almond scent that is released when the leaves are thoroughly crushed. The holly-leaf cherry reaches its northern distribution in the Bay Area; look for it on San Bruno and Montara mountains. It becomes increasingly plentiful as you go south, and large groves of sizeable trees are seen in many of the coastal mountains of south-central and southern California.

CHOKE-CHERRY *Prunus virginiana* var. *demissa*

Pls. 211, 212

The choke-cherry—a large, deciduous shrub or small tree—is nearly ubiquitous across the United States, growing in a wide range of different habitats and plant communities. The choke-cherry is characterized by lance-ovate, finely toothed

Plate 211. Choke-cherry. The choke-cherry bears rounded racemes of white, nectar-rich, plum-blossom-like flowers in late spring or summer.

leaves; narrow racemes of creamy white, plum-blossom-like flowers (pls. 211, 212); and deep purple drupes that ripen in late summer. Another diagnostic feature is the pair of tiny, wartlike glands at the base of each leaf; most prunuses have them. The fruit, true to the common name, is intensely bitter but makes an excellent jelly when cooked with sugar. The choke-cherry is only occasionally encountered in the Bay Area, and although it often finds its way onto lists of the local floras, there are seldom more than a few scattered plants in any one area. Look for the choke-cherry in shaded canyon

Plate 212. Choke-cherry. The choke-cherry is characterized by many leafy branches that terminate in slender racemes of white, plum-blossom-like flowers.

bottoms that receive some substantial sun for part of the day. It is localized on Mt. Diablo and along Mines Road south of Livermore. To appreciate its full potential, travel to the rivers that issue from the east side of the Sierra. Other deciduous prunuses in the Bay Area—the bitter cherry *(P. emarginata)* and Sierra plum *(P. subcordata)*—are also uncommon and seldom grow as trees.

Rutaceae (Rue or Citrus Family)

Mostly evergreen or deciduous shrubs or trees with variable, gland-dotted leaves having a pungent bitter-rue or citruslike odor. Clusters of often white, sometimes fragrant flowers have four or five sepals and petals, four to 10 stamens, and a pistil with a superior ovary. Fruits are thick-skinned berries (often called *hesperidiums*), capsules, or samaras. This mostly tropical family is noted for edible fruits in the genus *Citrus* (lemon, lime, orange, grapefruit, tangerine) and for a few garden ornamentals. California has three native genera, one of which attains small-tree status in the Bay Area.

HOPBUSH   *Ptelea crenulata*

Pls. 213–215

The hopbush gained its name from the use of the leaves as a substitute for hops in the brewing process, although when you smell the leaves, you'll find they have an aroma that is quite different from that of hops. Nonetheless, this odor—a sort of essence of bitter citrus—is a good identifying trait, for the dark glands that are responsible for the odor are scattered throughout the leaves. The leaves are trifoliate, with three broad, entire, pointed leaflets that are glossy when young (pl. 213), and so there is a certain superficial resemblance to

Plate 213. Hopbush. The hopbush has leaflets in threes and small clusters of whitish flowers reminiscent of *Choisya* (Mexican mock-orange) blossoms.

poison-oak *(Toxicodendron diversilobum)*, yet there are many differences when you look closely. Poison-oak leaflets, for example, are scalloped, rounded, and red when they first appear in spring. The hopbush, like so many other plants with the word *bush* in their name, is most often a large shrub, but it may attain small-tree status and grow up to 15 to 20 feet high. The multiple trunks fan out into twigs with reddish purple bark (pl. 214), which helps distinguish the hopbush when it's leafless in winter. After the leaves appear in early spring, they're followed by small clusters of creamy, four-petaled, starlike flowers (pl. 213) reminiscent of their distant Mexican cousin Mexican mock orange *(Choisya ternata)*. Round, single-seeded fruits surrounded by a translucent wing (pl.

Plate 214. Hopbush. The hopbush bears dark purple bark and a dense covering of shiny leaves.

Plate 215. Hopbush. The rounded, winged samaras of the hopbush linger after the leaves have dropped.

215) follow in summer. The hopbush is seldom abundant in the Bay Area, but for reasons that are not apparent, it is a common component of the chaparral on north-facing slopes and along streams on Mt. Diablo.

Salicaceae (Willow Family)

Deciduous shrubs or trees with alternate, stipuled, simple, unlobed leaves. Tiny, often wind-pollinated, petal-less flowers are borne in hanging or upright catkins, male and female on separate plants. Male flowers consist of a bract and two to several stamens; female flowers comprise a bract, a pistil, and a superior ovary. Fruits are capsules containing many hairy seeds that are wind dispersed. Both the cottonwoods (*Populus* spp.) and willows (*Salix* spp.) are well represented in California and occur in the Bay Area in riparian habitats. Cottonwoods are distinguished from willows by their hanging catkins and generally broader leaves. Willows have upright catkins and lance-shaped to linear leaves.

BLACK COTTONWOOD
Pls. 216–218

Populus balsamifera
subsp. *trichocarpa*

Cottonwoods are aptly named, for the female tree produces capsules filled with hair-covered seeds (pl. 216). These seeds are a miracle of packaging: somehow a large number of seeds

Plate 216. Black cottonwood. The black cottonwood is characterized by ovate, pointed leaves. Here you see the cottony masses of seeds emerging from the seed pods.

Plate 217. Black cottonwood. The dark bark of the black cottonwood is patterned with shallow, long, vertical fissures.

fits into a small space inside the capsules, yet when the seeds emerge, their cottonlike hairs expand to fill many times the original volume inside the capsule. This fluffy quality gives seeds great buoyancy on winds, but when the winds fail, great wads of cottony seeds drift to the ground. The black cottonwood is noted for its massive, fissured trunk with dark bark (pl. 217); its oval-shaped crown (p. 218); and its dark green, narrowly ovate leaves bordered by very fine saw teeth (pl. 216) or scallops. Like most other riparian trees, it grows rapidly to maturity, suckers abundantly, and reproduces prolifically. Both the pollen catkins and the seeds use wind for dispersal. Of the two Bay Area populuses, the black cottonwood is the less common, but it may be seen as a dominant tree among western sycamores *(Platanus racemosa)* and bigleaf maples

Plate 218. Black cottonwood. The black cottonwood quickly grows into an attractive tree with an oval-shaped crown.

(*Acer macrophyllum*) in Henry Cowell State Park along the San Lorenzo River, and at Hendy Woods State Park near Boonville in Mendocino County.

FREMONT COTTONWOOD   *Populus fremontii*

Pls. 219–221

The Fremont cottonwood lives in the same kinds of places the black cottonwood (*P. balsamifera* subsp. *trichocarpa*) does; both are widely distributed through many parts of California, and it's difficult to predict which species will live in a given area, except that the Fremont cottonwood is much more likely to be encountered in the Bay Area. The habit of the two trees is similar, as are their growth pattern and means of reproduction—including the male and female flower catkins (pl. 219), seed pods, and seeds—but the Fremont cottonwood has more deeply fissured, paler, brown gray bark (pl. 220) and broad, almost triangular leaves bordered by coarse scallops (pl. 221). Many riparian areas completely lack cottonwoods,

Top left: Plate 219. Fremont cottonwood. Trusses of female catkins emerge in early spring on the Fremont cottonwood. The male catkins appear on separate trees.

Above: Plate 220. Fremont cottonwood. The vertically fissured bark of the Fremont cottonwood is much paler than that of the black cottonwood.

Left: Plate 221. Fremont cottonwood. The Fremont cottonwood is identified by broadly ovate leaves edged with scalloped teeth. These branches are loaded with cottony seeds.

especially narrow streams with no discernible development of a floodplain. The larger streams and rivers feature them. Good cottonwood habitats include the Russian River, Mitchell Canyon on Mt. Diablo, and parts of Mines Road south of Livermore.

WILLOWS

Salix spp.

Pls. 222–228

The various willows constitute one of the most predictable and constant elements of riparian woodlands. One or more willows occur in this woodland, usually in the lower story, throughout California's riparian habitats. Few willows attain large-tree status and consequently are considered as large, multitrunked shrubs. I describe the various willow species found in the Bay Area under one heading because they share so many features: all are fast-growing, moisture-loving plants that root easily when their branches lean over into the mud. For this reason, willows are widely used to stabilize steep streambanks; an effective technique is to tie bundles of willow twigs together and bury them. Willows also feature pliant twigs that are easily bent into many shapes; the Indians used them to create frameworks for many of their baskets. Willow bark contains salicylic acid, the active ingredient in aspirin, making willows prominent in medicine. Willow twigs are typified by slender, pointed axillary buds that are pressed against the twig; simple lance-shaped to narrowly ovate, alternate leaves; and often collar-shaped stipules. All willows are deciduous and produce their flowers

Plate 222. Arroyo willow. The female catkins of willows are similar in shape and arrangement to male catkins, but they're green (the green part is the ovary, or future seed pod).

Plate 223. Arroyo willow. The male catkins of willows are stiffly upright and bear yellow stamens. The female catkins appear on separate trees.

from fuzzy "pussy willow buds" before new leaves appear in early spring. Female catkins consist of bracts and single, greenish pistils topped by sticky stigmas (pl. 222). Male trees produce upright male catkins consisting of many bracts and stamens (pl. 223). Although the design of the catkins and the lack of petals suggest wind pollination, willow flowers sometimes produce nectar and so attract bees. Willow flowers employ a double strategy: both insect and wind pollination. Willow ovaries ripen into capsules containing many hairy seeds, which when released are blown long distances by wind. (In this respect, willow seeds closely resemble cottonwood seeds.) The following willows are typical in the Bay Area: The weeping willow *(S. babylonica),* from the Near East, is occasionally naturalized and is widely grown as a graceful ornamental in landscapes receiving ample water. The weeping willow is easily told by its very broad crown and drooping branches that often touch the ground (pls. 224, 225)—its silhouette is unlike any other willow, although it might be confused from a distance with the Chilean maytens *(Maytenus boaria).* The arroyo willow *(S. lasiolepis)* is our most widespread native species, occurring as a bushy tree up to 25 feet high along watercourses throughout the foothills. Its few-toothed leaves are dark, dull green on top and whitish underneath (pl. 226). Two other tree willows

Right: Plate 224. Weeping willow. The nearly vertically hanging branches of the weeping willow leaf out in early spring.

Below: Plate 225. Weeping willow. Even in winter, when it's leafless, the shape of the weeping willow is unmistakable.

Plate 226. Arroyo willow. The undersides of arroyo willow leaves are pale and whitish. Here you can also see the characteristic lateral buds associated with the leaves.

Plate 227. Arroyo willow and red willow. The red willow stands taller and more treelike to the left of an arroyo willow. The arroyo willow generally has a bushy appearance.

occur in our area: the red willow *(S. laevigata),* a generally taller tree (compare the arroyo willow and the red willow in pl. 227), features red to yellowish twigs and leaves shaped much like those of the arroyo willow but with tiny scallops along the margins (pl. 228) and are whitish to bluish green underneath; the shining willow *(S. lucida* subsp. *lasiandra)* has brownish twigs and lance-shaped leaves that taper abruptly to a point with fine saw-toothed margins and a very shiny upper surface. The shining willow grows into a sub-

Plate 228. Red willow. The often drooping branches of the red willow bear narrowly lance-shaped leaves.

stantial tree, often reaching close to 50 feet high. Other features (look closely) that distinguish the three willows are that the red willow has buds covered with separate scales; the shining willow has buds with partly fused scales and pairs of tiny, wartlike glands at the base of the leaf blade; and the arroyo willow has buds with partly fused scales but no obvious glands on its leaf blades. Because sorting the various willows is difficult, these features should serve as useful guidelines. Finally, note that there are a few other willows in the Bay Area that never exhibit a dominant trunk and are always considered shrubs.

Simaroubaceae (Quassia Family)

Dioecious shrubs or trees with alternate, simple to pinnately compound leaves. They usually have racemes or panicles of small flowers with five erect sepals, five spreading petals, 10 to 15 stamens attached to a disk, and two to five separate pistils with superior ovaries. Fruits vary; they're sometimes winged samaras. This small family is best represented in tropical to warm temperate areas and is economically insignificant. California has one native desert shrub *(Castela emoryi)* and a widely naturalized tree.

CHINESE TREE-OF-HEAVEN *Ailanthus altissima*
Pls. 229–232 NOT NATIVE

Wherever it occurs, the Chinese tree-of-heaven is immediately recognized by the manner in which it grows in widely spaced groves of small trees (pl. 229). This tough tree is capable of producing prodigious numbers of suckers and can quickly colonize new areas. It was first brought into Califor-

Plate 229. Chinese tree-of-heaven. Rows of bare-twigged Chinese tree-of-heaven trees are often interconnected by a common, suckering root system.

Above: Plate 230. Chinese tree-of-heaven. The new leaves of the Chinese tree-of-heaven are initially flushed with red.

Right: Plate 231. Chinese tree-of-heaven. The tiny, whitish flowers of the Chinese tree-of-heaven emerge with the new leaves in spring.

Plate 232. Chinese tree-of-heaven. Here you see the pinnately compound leaves with drooping leaflets and the winged red fruits of the Chinese tree-of-heaven.

nia by indentured Chinese workers, and they planted it just about every place they lived. A strong testament to this practice is seen throughout the Gold Rush towns in the Sierra foothills, but many places in the Bay Area also feature these trees, and they often maintain themselves without anyone's help. Bare in winter and with grayish bark, the Chinese tree-of-heaven goes through many seasonal changes, starting with a flush of reddish new leaves in early spring (pl. 230). Each leaf quickly fills out into a long, pinnately compound structure consisting of several pointed and toothed leaflets that gently droop downward from their central midrib. Large panicles of tiny, whitish blossoms appear in early summer (pl. 231), followed by bright red, winged samaras by summer's end (pl. 232). The leaves finally turn brown in mid- to late fall, and the samaras fade to brown, leaving the trees bare once more.

Solanaceae (Potato or Nightshade Family)

Herbaceous plants, shrubs, or small trees, the trees usually have simple, unlobed, evergreen leaves that are often foul scented. Usually showy flowers have five partly fused sepals; five petals fused into a disk or tube, pleated in bud; five stamens; and a pistil with a superior, often two-chambered ovary. Fruits are several-seeded capsules or fleshy berries, like the tomato. This mostly subtropical and tropical family is especially diverse in the New World, and California has several species in various genera. The family is renowned for its deadly alkaloids (some have medicinal effects in minute doses) and several important food plants, including potato, tomato, eggplant, and chile and bell peppers. One introduced South American plant—the tree tobacco (*Nicotiana glauca*) —enters the Bay Area.

TREE TOBACCO *Nicotiana glauca*

Pl. 233   NOT NATIVE

The tree tobacco, a small, slender, gracefully drooping treelet, may grow up to 10 or 15 feet high but only sometimes develops a convincing trunk. It's included here because of its common name and the fact that it's now an increasingly prominent sight along dry, hot roadsides, where it flourishes without care. Introduced from Brazil and first naturalized in southern California, the tree tobacco appears to be enlarging its territory; it is now often seen in the inner Coast Ranges foothills, as, for example, along Mines Road. Although it is related to the tobacco plant (*N. tabacum*), the tree tobacco has little scent in its broadly elliptical, bluish, evergreen leaves. It is immediately recognized by the racemes of long, tubular, pale yellow flowers tipped by puckered and flared petal lobes (pl. 233). Because the flower tubes are full of nectar, hummingbirds are strongly drawn to them and help assure the

Plate 233. Tree tobacco. The tree tobacco is characterized by graceful branches, bluish green leaves, and long, tubular, yellow flowers.

success of seed production. The flowers appear through the warm months of the year.

Sterculiaceae (Cacao Family)

Herbaceous plants, shrubs, and rainforest trees, mostly with simple, sometimes palmately lobed, evergreen leaves. Leaves are often covered with starburstlike (stellate) hairs. The conspicuous flowers often have five partly fused, petal-like sepals (petals are often missing); five stamens fused into a tube; and a pistil with a superior, five-chambered ovary. Fruits are woody capsules. This modest family is mainly tropical and is famed for the cacao plant *(Theobroma cacao),* which is the source of chocolate. Two genera are native to California; flannel bush *(Fremontodendron californicum),* or California fremontia, sometimes reaches small-tree status in the Bay Area.

FLANNEL BUSH or CALIFORNIA FREMONTIA

Fremontodendron californicum

Pls. 234–236

The flannel bush is a common sight in the chaparral or along the edge of evergreen woodlands in southern California's mountains, especially in the southern Sierra foothills and in the Transverse Ranges. But it also occurs north into the central Sierra foothills and through the Coast Ranges as far as Napa and Sonoma counties, where it is mostly confined to serpentinite soils. The flannel bush may grow as a multi-branched shrub with many divaricating side branches, as a semiprostrate ground cover in a rare form that grows on Pinehill, or as a small tree up to 20 feet high. Its leaves and flowers also vary: typical leaves measure two to four inches across and are shallowly palmately lobed, but the forms found in the North Bay have much smaller leaves that often show little lobing. Flowers are typically golden yellow or yellow orange and resemble a webbed sea star (pl. 234), and flower size ranges from less than two inches to four inches across. Each flower consists of five, thick, leathery sepals and no petals, and

Plate 234. Flannel bush. The showy, saucer-shaped flowers of the flannel bush lack petals. Instead, the colorful sepals take their place.

five prominent, central stamens fused by their filaments into a white tube. The fruits ripen into five-angled, woody capsules covered with irritating hairs; in fact, most of the plant—young stems and leaves—is also covered with similar hairs, resulting in its common name. Look for the flannel bush along Hwy. 29 north of Mt. St. Helena, and in the upper part of Austin Creek State Park north of the Russian River. A small stand has also recently been documented in the Carson Ridge area of Marin County, and scattered populations are documented in the Santa Cruz Mountains.

Plate 235. Flannel bush. The flannel bush is smothered in late spring in yellow orange flowers. Often the trees are broader than tall.

Plate 236. Flannel bush. The flannel bush branches show a mixture of felted leaves and showy flowers.

Tamaricaceae (Tamarisk Family)

Evergreen shrubs or trees with jointed green twigs; tiny, scale-like leaves; and dense panicles of tiny, pink flowers. Flowers usually have five sepals, five petals, and five or 10 stamens attached to a lobed, nectar-secreting disk. The pistil has a superior ovary with two to five styles. Fruits are capsules with many hairy seeds. This small family is found in salty or alkaline areas from the Mediterranean region into Asia and Africa. Members of the genus *Tamarix* (tamarisks) have been widely planted in the hot areas of California, and occasionally they attain tree status in the Bay Area.

TAMARISK, SALT-CEDAR   *Tamarix* spp.
Pls. 237, 238   NOT NATIVE

Tamarisks have long been cultivated in hot areas, especially where soils are salty, to create fast-growing windbreaks and hedges, and to provide shade. But because their roots probe deeply for water, they often outcompete the native vegetation found in riparian areas and promote the salinity of the soils, with the result that they eliminate other plants. Tamarisks naturalize readily because of their hairy seeds, which carry well on winds. Most tamarisks look similar: they are large shrubs or small trees up to around 20 feet tall, with grooved bark; numerous slender, jointed, green twigs; and minute leaves reduced to tiny scales. When they burst into flower in spring, they create beautiful clouds of lavender pink (pl. 237), but the floral pageant doesn't last long. Each tiny flower may have four or five sepals and petals (*T. parviflora* has four; *T. gallica* has five), and a corresponding number of stamens attached to a nectar-bearing disk. You're most likely to encounter tamarisk in the "wild" along some of the creeks in the inner North Coast Ranges of Solano and Colusa counties (pl. 238).

Plate 237. Tamarisk. The scalelike leaves of the tamarisk are completely hidden by slender spikes of pink flowers in spring.

Plate 238. Tamarisk. The tamarisk may grow as a low, clumped shrub or as a tree, according to conditions. Often, the shrub form is what you see along Bay Area streams.

Taxaceae (Yew Family)

Evergreen, needle-bearing, coniferous shrubs or trees. Small, cream-colored pollen cones are borne on plants different from those that produce the fleshy, one-seeded seed cones. The seed cones are often referred to as "berries" and consist of either a fleshy cup around a seed or a plumlike wrapping surrounding the entire seed. This small, Northern Hemisphere family has two genera, both of which are found in California, but only the California-nutmeg *(Torreya californica)* is to be expected in the Bay Area. Some records indicate the Pacific yew *(Taxus brevifolia)* in one or two locales, but it is exceedingly rare here.

CALIFORNIA-NUTMEG   *Torreya californica*
Pls. 239–241

The California-nutmeg is a most unusual member of the conifer clan, for although it has needlelike leaves and typical pollen cones, the seed cones bear a single seed enclosed in a fleshy, plumlike wrapping (the aril) (pl. 239). The cone is the size and shape of a small prune and turns from pale green to purple by late fall. Inside is a large, nutmeg-shaped seed, which supposedly was gathered and offered for sale as the spice nutmeg to unsuspecting miners during the goldrush days. The true nutmeg *(Myristica fragrans)* is a tropical, evergreen, flowering tree and has no relationship whatsoever with the California-nutmeg. Only the female tree produces seed cones, and those are obvious only during part of the year, so the best identifying trait is the California-nutmeg's three- to four-inch-long glossy needles, which end in a spiny point (pl. 240). California's Indians used California-nutmeg needles for tattooing. The needles also carry an unusual fragrance, which may be pleasing or unpleasant according to your sense of smell; Willis Jepson found them unpleasant and dubbed this tree the "stinking yew." The California-nutmeg seldom grows

Plate 239. California-nutmeg. These California-nutmeg "fruits" (actually modified cones) are nearly ripe. When fully ripe they resemble small plums.

Plate 240. California-nutmeg. The white pollen cones of the California-nutmeg are borne among the glossy, wickedly sharp needles. Male cones occur on separate trees.

Plate 241. California-nutmeg. The California-nutmeg is seldom a large tree, but you can identify it by the stiff branches and drooping branchlets.

more than 50 or 60 feet tall, and its trunks rarely exceed two feet in diameter, so this is a small conifer by California standards. It is also noted for its drooping side branches (pl. 241). The California-nutmeg occurs in scattered groves with large gaps in its distribution, but there are several good stands within the Bay Area, including in Robert Louis Stevenson State Park on Mt. St. Helena, Samuel Taylor State Park, and the Bootjack Camp area on Mt. Tamalpais.

Taxodiaceae (Bald-cypress or Redwood Family)

Needle-bearing coniferous trees. The needles are shed attached to their twigs, unlike the needles in the pine family (except for pines, which shed their needles attached to minute spur shoots). Small pollen cones and seed cones are borne on the same tree. Seed cones consist of several spirally arranged, often diamond-shaped, woody scales that carry several seeds each. Most members of the family are noted for the lack of resin and pitch in their bark, whereas most pine relatives have an abundance of both. California is famed for its two redwoods: the giant sequoia *(Sequoiadendron giganteum),* with spirally arranged, prickly needles, and seed cones to four inches long; and the coast redwood *(Sequoia sempervirens),* with needles arranged in two rows (on the lower branches), and seed cones to about two inches long. Both are included in this book, because the giant sequoia sometimes grows in natural areas.

COAST REDWOOD   *Sequoia sempervirens*

Pls. 242–245

The coast redwood surely counts as one of the world's most famous trees. Confined to the fog belt of the Coast Ranges from Monterey County north into southernmost Oregon, the coast redwood may soar to over 360 feet tall or hunker down on coastal bluffs and remain only a few feet high, although this latter situation is highly unusual. Fast-growing and long-lived, the coast redwood lives in excess of 2,000 years, attains diameters of up to 18 feet, and is able to regenerate from stump sprouts. It's this ability to grow entirely new trees in circles around the parent tree that has permitted relatively rapid regeneration of logged-out redwood forests. (You will often see circles of trees where the parent trees have been logged out.) Although the coast redwood also produces abun-

Plate 242. Coast redwood. Coast redwood bark is a rich red brown and usually vertically furrowed.

dant crops of seed cones every year, the seeds seldom find a suitable home in the shade of the mature forest, but they germinate in large numbers after a flood or fire removes the original trees. The coast redwood grows best on floodplains a half mile or more back from the immediate coast, where it often occurs in nearly exclusive stands; the ability to sprout new roots after flooding gives it a competitive edge over other trees that lack this talent. Where redwoods grow shorter or less vigorously, as on steep slopes or after logging, the Douglas-fir *(Pseudotsuga menziesii),* California bay *(Umbellularia californica),* madrone *(Arbutus menziesii),* tanbark-oak *(Lithocarpus densiflorus),* red alder *(Alnus rubra),* and other trees grow successfully among them. Riparian corridors in redwood forest also feature red alder, bigleaf maple *(Acer macrophyllum),* cottonwoods *(Populus* spp.), and willows *(Salix* spp.). Redwoods were never extensive in the East Bay, and today the best stands occur in the tiny town of Canyon near Moraga, and in Redwood Regional Park. Although Muir Woods is a major tourist destination because of its redwoods, Big Basin State Park in the Santa Cruz Mountains and Armstrong Redwoods State Reserve north of Guerneville have aesthetically superior stands of this amazing tree. The coast redwood is best identified by its reddish brown, fibrous bark (pl. 242); the two-ranked needles on the lower branches (pl. 243); and the small

Plate 243. Coast redwood. The needles of the coast redwood lie flat on the lower branches. Here you see the contrast in color between the new flush of needles and last year's.

Plate 244. Coast redwood. The seed cones of the coast redwood are surprisingly small, with dark brown, wrinkled, diamond-shaped scales.

Plate 245. Coast redwood. The coast redwood bears branches that stick straight out from the trunk and often swoop up at their ends.

seed cones with wrinkled, diamond-shaped scales (pl. 244). It is also noted for the way the side branches diverge at right angles from the trunk (pl. 245). The upper branches differ in that they bear spirally arranged, prickly needles—people are often startled to discover these different-looking twigs on the ground after wind storms have brought them down.

GIANT SEQUOIA or BIGTREE   *Sequoiadendron*
Pls. 246, 247   *giganteum*

The giant sequoia—California's other redwood—is exclusive to the Sierra Nevada, where it occurs at middle elevations from the central Sierra south, growing in company with the sugar pine *(Pinus lambertiana),* incense-cedar *(Calocedrus decurrens),* white fir *(Abies concolor),* ponderosa pine *(P. ponderosa),* and others. I'm including the giant sequoia because it is sometimes seen in natural settings in the Bay Area, particularly in places where attempts at "aforestation," as in Tilden Regional Park, have included nonindigenous species. The giant sequoia is the bulkier, shorter brother of the coast redwood *(Sequoia sempervirens)*—what the latter achieves in height, the giant sequoia makes up for in bulk. Even so, the giant sequoia may grow to over 290 feet tall, but it really stands out for its enormous girth—up to 40 feet in diameter—which only gradually diminishes as the trunks pierce the air. Because of this enormous girth that extends so high, the giant sequoia claims the crown as the world's most massive tree. It is also counted among the longest lived, with individuals dated in excess of 3,000 years. The giant sequoia and the coast redwood differ sufficiently to be placed in separate genera: the giant sequoia has a stump that sprouts only when young; it grows in a symmetrical columnar shape when young (pl. 246); it has very thick, cinnamon-colored bark; its leaves are all awl shaped, sharp tipped, and pressed against the twigs (pl. 247); and its barrel-shaped seed cones reach over four inches long. These cones take nine months to ripen

Plate 246. Giant sequoia. Young giant sequoia trees are a perfect conical shape. This one is loaded with seed cones because of the stress of a
drought year.

Plate 247. Giant sequoia. The foliage of the giant sequoia consists of prickly, awn-shaped needles arrayed all around the twigs.

from pollination to seed set but fail to fall from the tree or to open unless singed by fire or damaged by the harvesting activities of the Douglas squirrel or by the deposition of beetle eggs in the cone scales. By contrast, a coast redwood stump sprouts abundantly and easily until it reaches approximately 400 years of age; it has thinner, red brown bark; its lower needles are arranged flat in two rows, whereas its upper needles are similar to those of the giant sequoia; and it produces small seed cones that reach about two inches long. Coast redwood seed cones ripen in one year, opening and falling as soon as they're ripe.

Ulmaceae (Elm Family)

Mostly deciduous shrubs and trees with simple, serrated leaves that have an oblique (off-center) base and deciduous stipules. Small clusters of greenish, petal-less flowers consist of four to six sepals, four to six stamens, and a pistil with a superior, one-chambered ovary topped by two styles. Fruits are small, fleshy drupes or nutlets surrounded by a circular wing. This small family occurs across the temperate regions of the Northern Hemisphere and also enters the tropics. It is noted for its ornamental street trees that belong to the genera *Ulmus* (elms) and *Zelkova.* California has two genera, a native hackberry *(Celtis reticulata)* and a few introduced and naturalized elms (*Ulmus* spp.). The elms are described below.

ELMS $\qquad$ *Ulmus* spp.

Pls. 248–250

To people from the eastern United States, the elm is an important deciduous shade tree and often lines streets in Midwestern towns. The tragedy there has been the demise of the American elm *(U. americana)* due to its susceptibility to Dutch elm disease. Elms in the Bay Area are also used as street trees, although not so extensively, and they have managed to sometimes escape from cultivation and live on their own in canyons and washes. As a group, elms are easy to recognize: the furrowed gray brown trunks are often circled with vigorous stump sprouts or suckers; the serrated, pinnately veined

Plate 248. Chinese elm. Elm leaves are coarsely serrate, are lopsided at the base, and have a strongly pinnate vein pattern.

Plate 249. Chinese elm. The greenish, wind-pollinated flowers of elms appear along the twigs while the trees are still leafless.

Plate 250. Chinese elm. Here you see the way the leaves are closely spaced along the branches of a large elm.

leaves display the characteristic off-center base (pl. 248); and the modest, green, wind-pollinated flowers (pl. 249) ripen into circular-winged fruits that are wind dispersed. The two elms most likely to be encountered here are the English elm *(U. minor)*, with doubly-serrated leaves whose undersides bear short hairs; and the Chinese elm *(U. parvifolia)*, with singly-serrated leaves that are basically hairless (pl. 250). Because of their capacity to sucker, elms may persist for a long time, for if the parent crown is damaged, the suckers grow into full-fledged trees on their own.

CHECKLIST OF
BAY AREA TREES

ACERACEAE (MAPLE FAMILY)
☐ *Acer macrophyllum*, bigleaf maple
☐ *Acer negundo* var. *californicum*, box elder

ANACARDIACEAE (SUMAC FAMILY)
☐ *Schinus molle*, Chilean-pepper tree

AQUIFOLIACEAE (HOLLY FAMILY)
☐ *Ilex aquifolium*, English holly

ARECACEAE (PALM FAMILY)
☐ *Phoenix canariensis*, Canary Island date palm

BETULACEAE (BIRCH FAMILY)
☐ *Alnus rhombifolia*, white alder
☐ *Alnus rubra*, red alder
☐ *Corylus cornuta* var. *californica*, California hazelnut

CAPRIFOLIACEAE (HONEYSUCKLE FAMILY)
☐ *Sambucus mexicana*, blue elderberry
☐ *Sambucus racemosa*, red elderberry

CELASTRACEAE (STAFF-TREE FAMILY)
☐ *Maytenus boaria*, maytens

CORNACEAE (DOGWOOD FAMILY)
☐ *Cornus nuttallii*, Pacific flowering dogwood

CUPRESSACEAE (CYPRESS FAMILY)

- ☐ *Calocedrus decurrens*, incense-cedar
- ☐ *Cupressus abramsiana*, Santa Cruz cypress
- ☐ *Cupressus goveniana* subsp. *goveniana*, Gowen cypress
- ☐ *Cupressus goveniana* subsp. *pygmaea*, pygmy cypress
- ☐ *Cupressus macnabiana*, Macnab cypress
- ☐ *Cupressus macrocarpa*, Monterey cypress
- ☐ *Cupressus sargentiana*, Sargent cypress
- ☐ *Juniperus californica*, California juniper

ELAEAGNACEAE (OLEASTER FAMILY)

- ☐ *Elaeagnus angustifolia*, Russian-olive

ERICACEAE (HEATHER FAMILY)

- ☐ *Arbutus menziesii*, madrone
- ☐ *Arctostaphylos glauca*, bigberry manzanita
- ☐ *Arctostaphylos manzanita*, common manzanita
- ☐ *Arctostaphylos regis-montana*, King Mountain manzanita
- ☐ *Rhododendron macrophyllum*, rosebay rhododendron

FABACEAE (PEA FAMILY)

- ☐ *Acacia dealbata*, silver wattle
- ☐ *Acacia longifolia*, Sydney golden wattle
- ☐ *Acacia melanoxylon*, blackwood
- ☐ *Albizia lophantha*, plume albizia or mimosa
- ☐ *Cercis occidentalis*, western redbud
- ☐ *Robinia pseudoacacia*, black locust

FAGACEAE (BEECH OR OAK FAMILY)

- ☐ *Chrysolepis chrysophylla*, coast chinquapin
- ☐ *Lithocarpus densiflorus*, tanbark-oak
- ☐ *Quercus agrifolia*, coast live oak
- ☐ *Quercus chrysolepis*, goldcup oak or canyon live oak
- ☐ *Quercus douglasii*, blue oak
- ☐ *Quercus garryana*, Garry oak or Oregon white oak
- ☐ *Quercus kelloggii*, California black oak
- ☐ *Quercus lobata*, valley oak or roble

- [ ] *Quercus parvula* var. *shrevei*, Shreve oak
- [ ] *Quercus wislizenii*, interior live oak

GARRYACEAE (GARRYA FAMILY)
- [ ] *Garrya elliptica*, coast silk-tassel

HIPPOCASTANACEAE (HORSE-CHESTNUT FAMILY)
- [ ] *Aesculus californica*, California buckeye

JUGLANDACEAE (WALNUT FAMILY)
- [ ] *Juglans californica* var. *hindsii*, northern black walnut
- [ ] *Juglans regia*, English walnut or Persian walnut

LAURACEAE (LAUREL FAMILY)
- [ ] *Umbellularia californica*, California bay or pepperwood or Oregon myrtlewood

MORACEAE (MULBERRY FAMILY)
- [ ] *Ficus carica*, edible fig

MYOPORACEAE (MYOPORUM FAMILY)
- [ ] *Myoporum laetum*, New Zealand myoporum

MYRICACEAE (SWEET-GALE FAMILY)
- [ ] *Myrica californica*, California wax-myrtle or bayberry

MYRTACEAE (MYRTLE FAMILY)
- [ ] *Eucalyptus camaldulensis*, river red gum
- [ ] *Eucalyptus globulus*, blue gum

OLEACEAE (OLIVE FAMILY)
- [ ] *Fraxinus dipetala*, flowering ash
- [ ] *Fraxinus latifolia*, Oregon ash
- [ ] *Olea europea*, olive

PINACEAE (PINE FAMILY)
- [ ] *Abies bracteata*, Santa Lucia fir
- [ ] *Abies grandis*, grand fir
- [ ] *Pinus attenuata*, knobcone pine
- [ ] *Pinus contorta* subsp. *contorta*, beach pine

- [ ] *Pinus contorta* subsp. *bolanderi*, Bolander pine
- [ ] *Pinus coulteri*, Coulter pine
- [ ] *Pinus lambertiana*, sugar pine
- [ ] *Pinus muricata*, bishop pine
- [ ] *Pinus ponderosa*, ponderosa pine
- [ ] *Pinus radiata*, Monterey pine
- [ ] *Pinus sabiniana*, gray pine or foothill pine
- [ ] *Pseudotsuga menziesii*, Douglas-fir
- [ ] *Tsuga heterophylla*, western hemlock

PITTOSPORACEAE (PITTOSPORUM FAMILY)

- [ ] *Pittosporum tobira*, tobira
- [ ] *Pittosporum undulatum*, Victorian box

PLATANACEAE (PLANE TREE FAMILY)

- [ ] *Platanus racemosa*, western sycamore

RHAMNACEAE (BUCKTHORN FAMILY)

- [ ] *Ceanothus oliganthus* var. *sorediatus*, jimbrush
- [ ] *Ceanothus thyrsiflorus*, blue blossom

ROSACEAE (ROSE FAMILY)

- [ ] *Amelanchier* spp., service berries
- [ ] *Cercocarpus betuloides*, mountain-mahogany
- [ ] *Cotoneaster* spp.
- [ ] *Crataegus suksdorfii*, western hawthorn
- [ ] *Heteromeles arbutifolia*, toyon or California-holly
- [ ] *Prunus cerasifera*, nonnative plum
- [ ] *Prunus ilicifolia*, holly-leaf cherry
- [ ] *Prunus virginiana* var. *demissa*, choke-cherry

RUTACEAE (RUE OR CITRUS FAMILY)

- [ ] *Ptelea crenulata*, hopbush

SALICACEAE (WILLOW FAMILY)

- [ ] *Populus balsamifera* subsp. *trichocarpa*, black cottonwood
- [ ] *Populus fremontii*, Fremont cottonwood
- [ ] *Salix babylonica*, weeping willow

- [ ] *Salix laevigata,* red willow
- [ ] *Salix lasiolepis,* arroyo willow
- [ ] *Salix lucida* subsp. *lasiandra,* shining willow

SIMAROUBACEAE (QUASSIA FAMILY)
- [ ] *Ailanthus altissima,* Chinese tree-of-heaven

SOLANACEAE (POTATO OR NIGHTSHADE FAMILY)
- [ ] *Nicotiana glauca,* tree tobacco

STERCULIACEAE (CACAO FAMILY)
- [ ] *Fremontodendron californicum,* flannel bush or California fremontia

TAMARICACEAE (TAMARISK FAMILY)
- [ ] *Tamarix gallica*
- [ ] *Tamarix parviflora*

TAXACEAE (YEW FAMILY)
- [ ] *Torreya californica,* California-nutmeg

TAXODIACEAE (BALD-CYPRESS OR REDWOOD FAMILY)
- [ ] *Sequoia sempervirens,* coast redwood
- [ ] *Sequoiadendron giganteum,* giant sequoia or bigtree

ULMACEAE (ELM FAMILY)
- [ ] *Ulmus minor,* English elm
- [ ] *Ulmus parvifolia,* Chinese elm

GLOSSARY

Accessory fruit A berrylike fruit such as a strawberry, whose fleshy receptacle engulfs several, tiny, one-seeded ovaries.

Achene A small, single-seeded ovary that does not open when ripe. Example: an unshelled sunflower "seed."

Alternate (leaf) One leaf per node along a stem.

Anther The sacs at the top of the stamen that produce pollen.

Arcuate (leaf veins) Veins that arch out from a leaf midrib.

Aril A fleshy, brightly colored wrapping around seed; attractive to birds.

Axil The upper angle between a leaf and its stem.

Berry A fleshy, many-seeded fruit. Examples: tomato, banana, blueberry, and grape.

Bisexual (flower) A flower that has both stamens (male part) and pistil (female part).

Bract Any modified leaf associated with a flower or flowers. Bracts may be green, like leaves, or colored, like petals. Examples: colorful bracts of the garden poinsettia, bougainvillea, and flowering dogwood.

Capsule A seed pod with two or more chambers that opens to shed its seeds.

Catkin A slender spike of tiny, greenish, brownish, or yellowish petal-less flowers. Examples: male flowers of hazelnuts and alders; male and female flowers of willows.

Compound (leaf) Any leaf composed of two or more separate leaflets. You can determine if it's a single leaf by looking for an axillary bud at the base. Compound leaves may be trifoliate, palmate, or pinnate.

Conifer A shrub or tree with needle- or scalelike leaves that bears its seeds in cones. Conifers don't produce true flowers. Examples: pines, firs, and cypresses.

Cyme A usually complex, multibranched, flat-topped flower cluster. Example: flowers of the blue elderberry.

Deciduous (plant) A plant that loses all its leaves before new leaves appear in spring.

Dioecious (plant) A plant that bears female flowers on a different plant from male flowers to assure cross-pollination. Most dioecious plants are wind pollinated. Examples: willows and cottonwoods.

Drupe A fleshy, single-seeded fruit, often with a single stone surrounding a central seed. Examples: avocado, date, cherry, and peach.

Elliptical A shape consisting of two converging half or part circles.

Entire (leaf, petal, or sepal) Without teeth, scallops, or lobes along the edges.

Evergreen A plant that retains at least some leaves year round. Evergreens are common among the conifers, but many broadleaved trees are also evergreen, such as live oaks, the madrone, and the California bay.

Genus A group of related species that share many traits. Example: all the different kinds of pines belong to the genus *Pinus*. Genus names are capitalized and italicized.

Gland A structure on leaves, stems, or flowers that secretes a sticky or watery substance. In leaves and hairs, glands produce sticky, smelly compounds to repel insects; in flowers, glands produce nectar to attract pollinators.

Habit The overall shape or form of a plant.

Habitat The environment in which a plant lives.

Head A dense, stemless cluster of flowers. Example: flowers of daisies, sunflowers, and asters.

Hypanthium A tube or cup on a flower that carries the sepals, petals, and stamens on top.

Inferior (ovary) An ovary positioned below other flower parts and clearly seen without pulling the flower apart. Examples: the ovaries in fuchsia and apple blossoms.

Leader   The main growing tip of a tree.

Leaflet   One individual part of a compound leaf.

Legume   A category of vegetables and the common name of a large plant family, and also a seed pod that has a single chamber and opens by two lengthwise slits. Examples: pea and bean seed pods.

Lobes   The indentations on a leaf or petal. Lobes may be arranged in a palmate or pinnate pattern.

Monoecious (plant)   A plant that bears male and female flowers on the same plant. Examples: oaks and alders.

Naturalized (plant)   A plant that appears to be growing naturally on its own without human assistance, rather than cultivated in gardens, parks, and orchards.

Needle   A leaf shaped like a needle. Examples: the leaves of pines and firs.

Nut   A ripe ovary that is surrounded by a hard shell with a large, usually edible seed inside. Examples: hazelnuts and the acorns of oaks.

Oblique (leaf)   A leaf with a lopsided base.

Opposite (leaf)   A leaf arranged with another, in pairs, on stems or twigs.

Ovary   The bottom saclike part of the pistil, which contains the future seeds or ovules. Ovaries ripen into various kinds of fruits as the seeds inside mature.

Ovate (leaf)   A leaf that is rounded at the base, gradually tapering to a pointed tip.

Palmate (leaf)   A leaf with lobes, veins, or leaflets that are arranged fanwise or like the fingers on a hand.

Panicle   A compound raceme of flowers, that is, a branch system that bears smaller side branches.

Perianth   The collective term for both sepals and petals.

Petal   The second row of parts in a typical flower. (Sepals form the outermost, or first, row.) Petals are usually colored to attract pollinators.

Petiole The stalk of a leaf.

Pinnate (leaf) A leaf with lobes, veins, and leaflets that are arranged like the parts of a feather.

Pistil The central, female part of a typical flower. The pistil consists of an ovary, a style, and a stigma.

Podzol A heavily leached whitish soil that is nearly devoid of nutrients.

Pollen The fine, often yellow powder produced by the stamens. Pollen needs to be deposited on a flower's stigma in order to grow a pollen tube and deliver its sperm to the egg inside the flower's ovules.

Pome A fleshy, many-seeded fruit such as an apple or pear in which the ovary itself is a papery sac surrounded by the succulent receptacle, or hypanthium.

Raceme Flowers borne along a stem on side branches.

Receptacle The end of the stem to which the flower parts are attached.

Relict A plant that once had a much broader distribution but is now restricted to special habitats because it has been outcompeted elsewhere.

Reticulate (leaf veins) Veins in a leaf that form a netlike pattern.

Samara A single-seeded fruit dispersed by an attached wing. Examples: the fruits of maples and ashes.

Scale A usually small bract or leaf that is shaped like a fish scale.

Sclerophyll (forest) Trees and shrubs with tough leaves adapted to relatively dry climates.

Sepal The outermost layer of a typical flower. Sepals are usually green and protect the flower in bud.

Serpentinite (soil) A soil derived from slick, greenish, bluish, or reddish metamorphic rocks that come from deep in the oceanic trench and have been changed by pressure and heat, then uplifted.

Simple (leaf) A leaf that is not divided into discrete, separate leaflets. Simple leaves may be lobed or toothed.

Species "Kinds" of plants. Although some species are variable, individual species are usually recognized as distinct and different from related species. For example, the oaks *(Quercus)* and pines *(Pinus)* have many different species: black oak, coast live oak, blue oak; gray pine, Monterey pine, sugar pine, and so on.

Spike A cluster of flowers borne directly along a stem.

Spur shoot A short, often stubby side branch borne along normal twigs or shoots. For example, pines produce clusters of leaves at the ends of tiny spur shoots; apples bear their flowers and fruits on specialized spur shoots.

Stamen The male part of a typical flower, consisting of a stalk (filament) and a sac (anther) and producing pollen.

Stellate (hair) A hair that is branched to resemble a starburst (use a good hand lens). Stellate hairs are typical of certain plant families, for example, the mallows (Malvaceae).

Stigma The tip of a flower's pistil, often sticky or fuzzy, where pollen is deposited during the process of pollination. Stigmas may be pinhead shaped, globular, or branched in various patterns.

Stipule A pair of leaflike appendages, often small, at the base of leaves. Certain plant families, such as the mallows (Malvaceae) and peas (Fabaceae), have stipules; others do not.

Stomatal band A stripe of a whitish waxy substance that surrounds the microscopic holes (stomates) of a leaf. Stomatal bands are conspicuous on the underside of certain conifer leaves such as those of the coast redwood and Santa Lucia fir.

Style A stalk between the ovary and stigma of the pistil.

Superior (ovary) An ovary positioned above the other flower parts.

Sweetpealike (flower) A flower like that of the pea family: the upper back petal stands up as the banner, the two outer petals protrude forward as wings, and the middle two petals come together as the boat-shaped keel.

Syconium A highly specialized fruit in which the tiny, single-seeded ovaries are borne along the inside of a hollow, fleshy receptacle. Example: the fig.

Thorn A spine-tipped side branch.

Trifoliate (leaf) A compound leaf that is divided into three leaflets. Examples: leaflets of the poison-oak and hopbush.

Two-lipped (flower) A flower in which two sets of petal lobes—usually two above and three below—are positioned at the end of a tube or throat. Two-lipped flowers are typical of the mint family (Lamiaceae) and snapdragon family (Scrophulariaceae).

Umbel A cluster of flowers arranged in an umbrella-like fashion, where all the flower-bearing stems radiate out from the end of a single large stem. Examples: flowers of members of the parsley family (Apiaceae), such as carrot and celery.

Umbo The enlarged end of the scale of conifer cones.

Unisexual (flower) A flower that bears only stamens (male) or only pistils (female).

Whorl Three or more leaves or flowers attached to the same level of a stem. Examples: lily leaves and flowers in the mint family (Lamiaceae).

FURTHER READING

This short list gives the reader more information on our wonderful native trees. These books all extend beyond the confines of the San Francisco Bay Region but cover many of the trees discussed in this book. Some books are no longer in print but can be found on the Internet or in used bookstores and are well worth the search. Three of the current books—*Conifers of California, The Oaks of California*, and *Trees and Shrubs of California*—do an excellent job of covering the trees found in the San Francisco Bay Region.

Arno, Stephen F. 1973. *Discovering Sierra trees.* Yosemite Natural History Association and Sequoia Natural History Association in cooperation with the National Park Service. A classic paperback describing the trees of the Sierra Nevada, some of which are found in the Bay Area. There is a center spread showing distribution of trees by elevation and slope. The illustrator, Jane Gyer, has done surpassingly fine scratchboard renderings of the trees.

Berry, James Berthold. 1966. *Western forest trees.* New York: Dover Publications. The selection in this book is for foresters, but it has incredibly detailed line drawings.

Griffin, James R., and William B. Critchfield. 1972. *The distribution of forest trees in California.* Pacific Southwest Forest and Range Experiment Station, USDA Forest Service Research Paper PSW-82/1972. Washington, D.C.: U.S. Government Printing Office. Written descriptions as well as distribution maps for all the native trees in California.

Hickman, James C., ed., 1993. *The Jepson Manual.* Berkeley: University of California Press. A wonderful book with keys to California's native trees and some fine line drawings.

Keator, Glenn. 1998. *The life of an oak: An intimate portrait.* Berkeley: Heyday Books and the California Oak Foundation. Although the book does not deal exclusively with native oaks, it introduces topics concerning oak reproduction, taxonomy, evolution, ecology, and more. Lavishly illustrated with color photos and watercolor paintings.

Lanner, Ronald M. 1999. *Conifers of California.* Los Olivos, Calif.: Cachuma Press. An expertly written book covering all of California's native conifers, with lovely watercolor paintings, photographs, and distribution maps.

McMinn, Howard E., and Evelyn Maino. 1981. *An illustrated manual of Pacific Coast trees.* Berkeley: University of California Press. Another classic book on trees, covering the entire Pacific coast. The book contains keys, descriptions, suggested uses, and some simple line drawings.

Pavlik, Bruce, Pamela Muick, Sharon Johnson, and Marjorie Popper. *The oaks of California.* Los Olivos, Calif.: Cachuma Press and The California Oak Foundation. A wonderful book detailing most of California's oaks, with keys, photos, descriptions, discussions of ecology, and remarks on preservation.

Peattie, Donald Culross. 1953. *A natural history of western trees.* New York: Bonanza Books. Although this large tome is not all-inclusive, it contains most of the prominent trees in the entire western United States. Besides having superbly rendered black-and-white illustrations by Paul Landacre, the book conveys a real sense of the trees through the author's prose.

Stuart, John D., and John O. Sawyer. 2001. *Trees and shrubs of California.* Berkeley: University of California Press. An excellent guide to the major woody plants of California, including most of the native trees and the most prominent native shrubs. The distribution maps and fine line drawings are by Andrea Pickert.

INDEX

Page references in **boldface** refer to the main discussion of the species.

ABOUT THE AUTHOR

Glenn Keator is a Bay Area botanist and writer who received his Ph.D. at the University of California, Berkeley. He currently resides in Berkeley. He has taught classes and led field trips for the California Academy of Sciences, Strybing Arboretum Society, Point Reyes Field Seminars, and Friends of Regional Parks Botanic Garden and is also a part-time instructor at Merritt College in Oakland. He is the author of *Complete Garden Guide to Native Perennials of California, Complete Garden Guide to Native Shrubs of California, Plants of the East Bay Parks,* and *The Life of an Oak: An Intimate Portrait.*

Series Design:	Barbara Jellow
Design Enhancements:	Beth Hansen
Design Development:	Jane Tenenbaum
Cartographer:	Bill Nelson
Composition:	Impressions Book and Journal Services, Inc.
Text:	9.5/12 Minion
Display:	Franklin Gothic Book and Demi
Printer and binder:	Everbest Printing Company